Building Better Models with JMP® Pro

Jim Grayson · Sam Gardner · Mia L. Stephens

support.sas.com/bookstore

The correct bibliographic citation for this manual is as follows: Grayson, Jim, Gardner, Sam, and Stephens, Mia. 2015. *Building Better Models with JMP® Pro.* Cary, NC: SAS Institute Inc.

Building Better Models with JMP® Pro

Contents

Part 4 Model Selection and Advanced Methods 221

Chapter 8 Using Cross-Validation 223

viii

Acknowledgments

Dedication

To all the people who have patiently encouraged and supported me through this project, especially my wife and children.

--SG

To Michael, Muffin, and Bubba – thanks for your patience and support!

--MS

To my parents, John and Beverly, and my wife, Cindy – my great encouragers.

--JG

Acknowledgments

We wish to express our gratitude to the reviewers who provided many helpful suggestions. This book is much better because of your contributions.

Thanks to:

Michele Boulanger
Michael Crotty
Goutam Chakraborty
William Duckworth
Suneel Grover
Duane Hayes
Matt Liberatore
Robert Nydick
Dan Obermiller
Sue Walsh

x

About This Book

Purpose

This book is designed for the student wanting to prepare for their professional career who recognizes the need for understanding of both the mechanics and the concepts of predominant analytic modeling tools and also how to apply methods in solving real-world business problems. This book is also designed to meet the needs of the practitioner who wants to obtain a hands-on understanding of business analytics to make better decisions from data and models and to apply these concepts and tools to business analytics projects.

Is This Book for You?

This book is for you if you want to explore the use of analytics for making better business decisions, and have either been intimidated by books that focus on the technical details or discouraged by books that primarily focus on the high level importance of using data without getting to the "how to" of the methods and analysis.

Prerequisites

While not required, it would be very helpful if the reader has taken a basic course in statistics. Experience with the book software, JMP Pro, is not required.

Scope of This Book

This book is not an introductory statistics book. While we provide an introduction to basic data analysis, data visualization and analyzing multivariate data, for the most part, the statistical details and background information are not provided. This book is also not a highly technical book that dives deeply into the theory or algorithms, but it will provide insight into the "black box" of the methods covered.

This book will cover analytic topics including:

- Regression
- Logistic regression
- Classification and regression trees
- Neural networks
- Model cross-validation

It will also provide an introduction to some advanced modeling techniques (boosting, bagging, and regularization) in an example driven structure.

About the Examples

Software Used to Develop the Book's Content

JMP Pro 12 is the software used throughout this book.

Example Code and Data

You can access the example code and data for this book by linking to its author page at http://support.sas.com/publishing/authors. Select the name of the author. Then, look for the cover thumbnail of this book, and select Example Code and Data to display the SAS programs that are included in this book.

Some resources, such as instructor resources and add-ins used in the book, will be posted on the JMP community file exchange (community.jmp.com).

For an alphabetical listing of all books for which example code and data is available, see http://support.sas.com/bookcode. Select a title to display the book's example code.

If you are unable to access the code through the Web site, send e-mail to saspress@sas.com.

Exercise Solutions

We strongly believe that for the reader to obtain maximum benefit from this book they should "play along" and complete the examples demonstrated in each chapter. Also, at the end of each chapter are suggested exercises to practice what has been discussed in the chapter.

Additional Help

Although this book illustrates many analyses regularly performed in businesses across industries, questions specific to your aims and issues may arise. To fully support you, SAS Institute and SAS Press offer you the following help resources:

- For questions about topics covered in this book, contact the author through SAS Press:
    ○ Send questions by email to saspress@sas.com; include the book title in your correspondence.
    ○ Submit feedback on the author's page at http://support.sas.com/author_feedback.
- For questions about topics in or beyond the scope of this book, post queries to the relevant SAS Support Communities at https://communities.sas.com/welcome.
- SAS Institute maintains a comprehensive website with up-to-date information. One page that is particularly useful to both the novice and the seasoned SAS user is its Knowledge Base. Search for relevant notes in the "Samples and SAS Notes" section of the Knowledge Base at http://support.sas.com/resources.
- Registered SAS users or their organizations can access SAS Customer Support at http://support.sas.com. Here you can pose specific questions to SAS Customer Support; under *Support*, click *Submit a Problem*. You will need to provide an email address to which replies can be sent, identify your organization, and provide a customer site number or license information. This information can be found in your SAS logs.

Keep in Touch

We look forward to hearing from you. We invite questions, comments, and concerns. If you want to contact us about a specific book, please include the book title in your correspondence.

Contact the Author through SAS Press

- By e-mail: saspress@sas.com
- Via the Web: http://support.sas.com/author_feedback

Purchase SAS Books

For a complete list of books available through SAS, visit sas.com/store/books.

- Phone: 1-800-727-0025
- E-mail: sasbook@sas.com

Subscribe to the SAS Training and Book Report

Receive up-to-date information about SAS training, certification, and publications via email by subscribing to the SAS Training & Book Report monthly eNewsletter. Read the archives and subscribe today at http://support.sas.com/community/newsletters/training!

Publish with SAS

SAS is recruiting authors! Are you interested in writing a book? Visit http://support.sas.com/saspress for more information.

About These Authors

Jim Grayson, PhD, is a Professor of Management Science and Operations Management in the Hull College of Business Administration at Georgia Regents University. He currently teaches undergraduate and MBA courses in operations management and business analytics. Previously, Jim held managerial positions at Texas Instruments in quality and reliability assurance, supplier and subcontractor management, and software quality. He has a PhD in management science with an information systems minor from the University of North Texas, an MBA in marketing from the University of North Texas, and a BS from the United States Military Academy at West Point.

Sam Gardner is a Senior Research Scientist at Elanco, a business division of Eli Lilly and Company. He currently works as a statistician supporting manufacturing development for a variety of animal health products. He is recognized by the American Statistical Association as an Accredited Professional Statistician. He has an MS in mathematics from Creighton University and an MS in statistics from the University of Kentucky. He graduated from Purdue University with BS degrees in mathematics and chemistry.

Gardner started his professional career as a military officer in the US Air Force, where he had roles that focused on modeling and simulation, operational flight test planning and analysis, and research and development. He also taught statistics at the Air Force Institute of Technology. After leaving the military, he began his work in the pharmaceutical industry as a statistician supporting the development and manufacturing of active pharmaceutical ingredients, and later transitioned to a role as a chemist in pharmaceutical manufacturing. He also worked as a statistician in pharmaceutical marketing with a focus on using statistical modeling to help solve business problems related to sales and marketing effectiveness. An avid user of statistical software, he spent several years working for JMP®, a division of SAS®, where he worked as a product expert and seminar speaker, with a focus on data visualization, applied statistics, and modern statistical modeling.

 Mia L. Stephens is an Academic Ambassador with JMP®, a division of SAS®. Prior to joining SAS®, she split her time between industrial consulting and teaching statistics at the University of New Hampshire. A coauthor of *Visual Six Sigma: Making Data Analysis Lean* and *JMP® Start Statistics: A Guide to Statistics and Data Analysis Using JMP®, Fifth Edition*, she has developed courses and training materials, taught, and consulted within a variety of manufacturing and service industries. Stephens holds an MS in statistics from the University of New Hampshire and is located in York Harbor, Maine.

Learn more about these authors by visiting their author pages, where you can download free book excerpts, access example code and data, read the latest reviews, get updates, and more:

http://support.sas.com/grayson
http://support.sas.com/gardner
http://support.sas.com/stephens

P a r t 1

Introduction

Chapter 1 Introduction

Chapter 2 Model Building and the Business Analytics Process

Part I discusses the need in business for analytical thinking and a broad understanding of analytical tools and techniques. It discusses the objectives of this book, presents an overview of exploratory and predictive modeling, and introduces the Business Analytics Process (or BAP).

1 Introduction

Overview

The words "analytics", "business analytics," and "data analytics" have become common-place in our society. Whether you are reading the *Wall Street Journal* or listening to a sports talk show, you may hear the word "analytics." We believe that to some extent every professional employee in every organization will interact with analytics in some form. Although the tools and methods used can get fairly technical, at its core analytics is about making better decisions with data.

We have written this book to provide an accessible, understandable, and hands-on introduction to this rich subject. Our focus is on interacting with data and building statistical models to enable better decision-making.

Analytics Is Hot!

Thomas Davenport, a well-known author and thought leader, describes analytics in *The New World of Business Analytics* (March 2010) as:

> …"business analytics" can be defined as the broad use of data and quantitative analysis for decision-making within organizations. It encompasses query and reporting, but aspires to greater levels of mathematical sophistication. It includes analytics, of course, but involves harnessing them to meet defined business objectives. Business analytics empowers people in the organization to make better decisions, improve processes, and achieve desired outcomes. It brings together the best of data management, analytic methods, and the presentation of results—all in a closed-loop cycle for continuous learning and improvement.

Better decisions and improved processes enable organizations to operate more efficiently (saving money) and to become more effective (better outcomes). There is a great demand for analytic talent that can help companies achieve these results.

A **McKinsey Global Institute** study identified a growing need for deep analytic talent. Universities have responded to this need. In 2007, North Carolina State University established the first Master of Science in Analytics (MSA) degree. Eight years later, there are more than 70 programs that offer graduate degrees in analytics or data science as shown in the chart below (Figure 1.1, from North Carolina State University Institute for Advanced Analytics, http://analytics.ncsu.edu/).

Figure 1.1: Growth in Graduate Degree Programs (Source: NCSU)

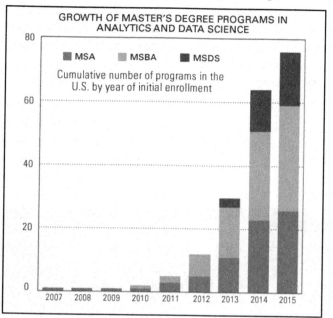

These graduate programs prepare individuals to become what is typically called a *data scientist*, but companies have an even greater need for managers and analysts who understand analytics. As the McKinsey study further states:

> In addition, **we project a need for 1.5 million additional managers and analysts** [authors' emphases] in the United States who can ask the right questions and consume the results of the analysis of big data effectively. The United States—and other economies facing similar shortages—cannot fill this gap simply by changing graduate requirements and waiting for people to graduate with more skills or by importing talent (although these could be important actions to take). It will be necessary to retrain a significant amount of the talent in place; fortunately, this level of training does not require years of dedicated study.

This book was written for these "managers and analysts," and for any students or professionals who need to better understand how to use data and models to make sound business decisions.

What You Will Learn

Business analytics is not monolithic. Rather, it encompasses three key and somewhat discrete categories: **descriptive**, **predictive**, and **prescriptive** analytics. Descriptive analytics describes *what* is happening, predictive analytics determines *why* it is happening and why it is likely to happen, and prescriptive analytics prescribes the *best action* to take.

These categories are described in Figure 1.2. This figure is from the International Institute for Analytics (IIA), which was adapted from *Competing on Analytics* (Davenport and Harris, 2007).

Figure 1.2: Analytic Methods

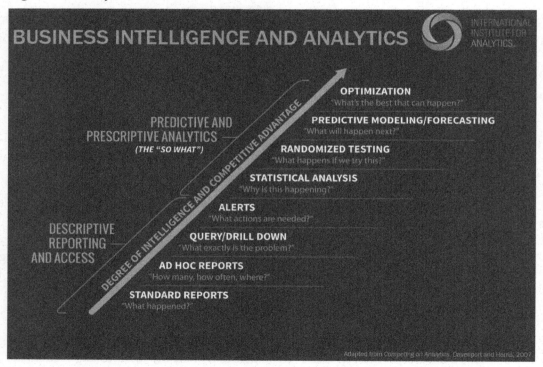

In this book, we focus on the "statistical analysis" and "predictive modeling" elements that are shown in Figure 1.2. Statistical analysis includes both *exploratory analysis* and *exploratory modeling*. In exploratory analysis, the goal is to become familiar with the data and to gain insights into the data structure and the variables involved. In exploratory modeling, the goal is to understand potential relationships between variables and to identify the most important variables. The purpose of *predictive modeling* is to predict new observations—to determine what is likely to happen in the future given the current process and business environment.

Analytics and Data Mining

There are many different flavors or styles of analytics—data analytics, marketing analytics, web analytics, and business analytics to name a few. The particular approach used depends largely on the field or application area. In this book, the focus is on *business analytics*. Namely, we focus on the application of the *Business Analytics Process* and analytic tools to business-oriented problems and opportunities.

Another popular process is *data mining*. Although both terms are often used interchangeably, business analytics and data mining are not the same. The differences are somewhat subtle. Data mining, according Linoff and Berry (2011) is "a business process for exploring large amounts of data to discover meaningful patterns and rules." The focus in data mining is on identifying hidden patterns and relationships, using methods such as machine learning, artificial intelligence, and statistical tools. Analytics, in general, is much broader. In analytics, data mining tools are used to find patterns and understand potential relationships, where the focus is on explaining why particular results occurred, understanding what might happen in the future, and applying what is learned within the context of the business problem or opportunity.

How the Book Is Organized

This book is organized in four parts:

- Part I: Introduction
- Part II: Preparing for Modeling
- Part III: Model Building
- Part IV: Model Selection and Advanced Methods.

In the introductory chapters (**Part I**), we provide an overview of model building and the business analytics process.

In **Part II, Preparing for Modeling (Chapter 3, Working with Data)** we provide an introduction to basic navigation and use of JMP. We cover a variety of tools for data visualization, exploration, and basic statistical analysis. Finally, we discuss some common issues with data quality and introduce some tools for data preparation, an essential step before beginning the model building process.

In **Part III, Model Building,** we introduce four foundational modeling methods: **Multiple Linear Regression (Chapter 4), Logistic Regression (Chapter 5), Decision Trees (Chapter 6),** and **Neural Networks (Chapter 7).** In each chapter, we identify business use cases for the particular modeling method, take a look "under the hood" at some of the technical details behind the method, and provide two case studies involving application of the method to a business problem. Each chapter also includes a number of exercises.

Lastly, in **Part IV, Model Selection and Advanced Methods,** we formally introduce methods for validation of predictive models (**Cross-Validation, Chapter 8**), and revisit examples introduced in previous chapters. In **Advanced Methods (Chapter 9),** we introduce some advanced model-building tools and techniques. We conclude with **Capstone and New Case Studies (Chapter 10).** In this chapter, we revisit the Business Analytics Process, applying the entire process to a new case study, and we introduce new examples based on large and messy data sets that are more representative of real business problems.

Let's Get Started

This book will walk you through the "what, why, and how" of analytic modeling methods. We believe it is important to be a "hands-on" learner. Download a trial version of the JMP software and follow along with us as we show you how to build better models. We've provided JMP menus, instructions, and keystrokes where needed to guide you. Now, let's get started.

JMP and JMP Pro are used throughout this book. JMP Pro includes an advanced set of tools for predictive modeling not available in JMP. In Chapters 3 through 7, we primarily use the standard version of JMP, which includes all of the tools for data visualization, analysis, and modeling that are introduced in these chapters. In Chapters 8 through 10, we use some of the advanced modeling features available only in JMP Pro. For a trial version of JMP, visit jmp.com/trial. JMP and JMP Pro may also be licensed through your school or organization. Check with your software administrator for availability and download information.

References

Davenport, Thomas H., and Jeanne G. Harris. 2013. *Competing on Analytics, The New Science of Winning*. Harvard Business Review Press. http://www.sas.com/content/dam/SAS/en_us/doc/event/The-Era-of-Impact-127837.pdf

Linoff, Gordon, and M. Michael Berry. 2011. *Data Mining Techniques: For Marketing, Sales, and Customer Relationship Management*, 3rd ed., Wiley. Chapter 1.

Shmueli, Galit. 2010. "To Explain or to Predict?" *Statistical Science*, Vol. 25, No. 3, 289-310.

Shmueli, Galit, Nitin R. Patel, and Peter C. Bruce. 2010. *Data Mining for Business Intelligence: Concepts, Techniques, and Applications in Microsoft Ofice Excel with XLMiner*, 2nd ed. John Wiley & Sons, Inc.

An Overview of the Business Analytics Process

Introduction

In this chapter, we describe a number of approaches for managing data mining and analytics projects, and introduce a methodology that we refer to as the Business Analytics Process (BAP). We walk through the key steps in this process, along with the core activities completed within each step. In later chapters we revisit these steps and introduce concepts and techniques used throughout the process.

Commonly Used Process Models

In a business setting, analytics projects can be complex, involving large amounts of data and stakeholders from various parts of the organization. Having a common framework for the analytics process is instrumental to project success. For data-mining oriented projects, two popular and well-documented processes are **SEMMA** and **CRISP-DM**:

- **SEMMA** is a data mining process developed by the SAS Institute, which stands for <u>S</u>ample, <u>E</u>xplore, <u>M</u>odify, <u>M</u>odel, and <u>A</u>ssess. While SEMMA was designed to be used in conjunction with the SAS system, it can be viewed as a general process for developing statistical models.

- **CRISP-DM**, which stands for <u>Cr</u>oss <u>I</u>ndustry <u>S</u>tandard <u>P</u>rocess for <u>D</u>ata <u>M</u>ining, was developed by a consortium of SPSS, NCR and Teradata, Daimler AG, and OHRA. CRISP-DM was designed to be general enough to use in any industry and is not tied to any specific tool or application. The major phases in CRISP are Business Understanding, Data Understanding, Data Preparation, Modeling, Evaluation, and Deployment (Chapman, 2000).

Gordon Linoff and Michael Berry in their book, *Data Mining Techniques*, 3rd ed. (2011), describe data mining as a four-stage business process, which they refer to as a "virtuous cycle": Transform data, act on the information, measure the results, and identify business opportunities.

The professional organization INFORMS (Institute for Operations Research and the Management Sciences) has developed an analytics certification, the **Certified Analytics Professional**, or CAP. The certification exam is built around the Analytics Job Task Analysis (http://bit.ly/1z850XW), which provides an outline of typical tasks performed by analytics professionals. However, this list of tasks does not constitute an "analytics process" per se.

Examining these data mining processes, along with the CAP job task analysis categories, reveals some common elements. None of the approaches is linear. Each allows for looping back to previous steps as new insights are discovered. They are also iterative — new business problems are identified and the process begins again. Finally, each process contains elements of collecting and handling data, performing statistical modeling, and applying the resulting models to solve a business problem.

The Business Analytics Process

Combining the best aspects of these processes, the authors propose an approach that we call the *Business Analytics Process* (BAP). The BAP steps are shown in Figure 2.1 and are described below.

Figure 2.1:   Business Analytics Process

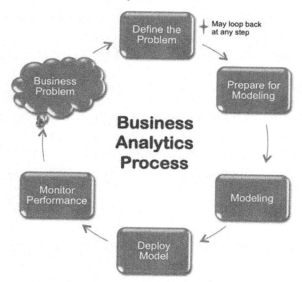

Define the Problem

The key outcome of this step is that the problem and the project are well-defined. Some of the main activities include:

- Understanding the business problem (or simply, the problem), project objectives, and importance to the organization
- Framing the analytics problem(s)
- Defining the project goal and time frame
- Developing a project plan and time line
- Obtaining resources and approval to start project

The BAP starts with clearly stating the business problem that will be addressed and then translating this problem into an analytics problem. The business problem is defined by the project sponsor or champion, in conjunction with management and key business stakeholders. Members of the team that will tackle the problem are identified, and the analytics team is formed. The sponsor, along with the analytics team, then reframes the business problem into an actionable analytics problem. Implicit in this problem formulation is that the team can define measurable responses that are related to the desired outcome or the observed problem. For example, a company may need to increase its customer base while growing profits. This is the business problem. To help accomplish this, we (the analytics team) may want to understand behaviors and characterize the demographics of current customers and build models that predict the future profitability of new and existing customers (the analytics problem). Ultimately, this information may be used to develop a new advertising campaign or targeted marketing programs.

The sponsor will continue to provide direction, support, and accountability during the entire BAP. Ultimately, the sponsor is the person(s) who says, "This is an important problem to solve." The sponsor then has the authority to provide the necessary resources to help find and act upon the solution. It is also important that the sponsor has an understanding of the BAP and supports this disciplined and structured approach to problem solving.

Prepare for Modeling

This step is all about compiling data and preparing data for analysis and modeling. Key activities are:

- Collecting, cleaning, and transforming the data
- Defining relevant features in the data
- Examining and understanding the data
- Producing data sets that are ready for analysis and model-building

A well-defined problem allows us to obtain appropriate data and prepare the data for modeling. Data must be collected and explored before building models. In many cases, it must also be cleaned, transformed and/or restructured. This step includes examining the data and developing insights, or tentative hypotheses, regarding the drivers of the problem that we want to solve. Take, for example, the analytics problem of predicting customer profitability. It is likely that corporate databases of customer transactions and customer profiles will be extracted, cleaned, and merged. Other information, such as regional, national, and global economic factors, competitor sales, and advertising can also be incorporated into the data used for modeling.

This part of the process is often the most time consuming. It is important that the team documents the data sources and the steps to take to compile and prepare the data. This includes knowing how the data is generated, collected, stored, queried, manipulated, and joined. This "data pedigree" (Hoerl et al., 2014) is an essential part of the analytics process. Common problems with data and core tools for data preparation are covered in Chapter 3.

Modeling

The end result of this step is a model, or set of models, that addresses our problem. In this step, the team:

- Chooses the appropriate modeling methods
- Fits one or more models
- Evaluates the performance of each model
- Chooses the best model or set of models to address the analytics problem (and ultimately the business problem)

This book is fundamentally about building models. But, what do we mean by the term "model" or "statistical model"? Statistical models describe how variables are related to one another. They allow us to predict or explain the future behavior of a process or system as a function of past behavior. Many different analytic methods may be used in search of the best model (or best combination of models) to predict the outcome(s) of interest. It is not uncommon to loop back to a previous step as we gain insight and identify additional data or features that are needed. We may also find that the original problem definition was inadequate and that the project plan needs to be revised.

The completion of this step leads to a set of models that addresses the analytic problem(s) we are trying to solve. Different models may be required to answer the different questions posed in the problem definition. For instance, to predict customer profitability we may use one model to predict the probability for customers that make purchases from our company and a separate model to predict the particular product(s) that that customer purchases and the profits associated with those purchases.

Deploy Model

This step is about putting the model (or models) into use. The analytics team:

- Delivers the model and model results to the business partners or internal customers

- Assists in applying model insights and implementing ongoing use of the model
- Documents the project
- Follows up with the business sponsor to close out the project

The final result of the process is to deploy the model for use within the organization. Sometimes the "deployment" is simply a matter of documenting the results of the modeling effort and the recommended improvements or changes that will solve the business problem. In other situations, the model that is developed will be integrated into the decision-making process for a particular part of the business. In this case, model deployment often relies on other business partners or systems, such as IT or Engineering. It is often wise to include these areas as stakeholders in the **Define the Problem** step. At the very least, the project sponsor should have the necessary influence or authority to ensure these resources are available.

Monitor Performance

Ongoing monitoring ensures that the model(s) continues to produce the desired results. Primary activities include:

- Monitoring model performance and refining the model if needed
- Evaluating and quantifying the improvement realized by the business as a result of the changes or solutions that were implemented.
- Determining additional business analytics problems to be solved

Again, this step depends on how the chosen model was used to solve the business problem. If the result is a recommendation for new business policies or process settings to solve a particular problem or achieve a business objective, then some sort of check or follow-up is needed to ensure that these changes are actually being followed, and that the model's predicted outcome has indeed been achieved. If the model is being used as an ongoing business tool, then monitoring the model performance may require ongoing recording of what was predicted and what actually occurred for each decision made, and monitoring to see if the predictions match reality. If it is found that that model is no longer performing as desired, then, in essence, a new business problem has been found, and this may lead to going back to the **Define the Problem** step.

Conclusion

In this chapter, we introduced the Business Analytics Process (BAP). The BAP is an approach to modeling that incorporates the best features of most data mining processes and analytics tasks and is well suited for users of JMP Pro in a variety of applications and industries.

In the next chapter, we provide an introduction to basic navigation and use of JMP and introduce tools used in the **Prepare for Modeling** step. In Chapters 4 through 9, different modeling techniques will be highlighted using a variety of examples, and we'll take a peek under the hood for each of these methods. Each chapter will include an example that is relatively straightforward and "textbook," and most of the chapters have examples that are more detailed, comprehensive, and typical of a real-life modeling situation. In these chapters, we'll focus on the first three steps of the Business Analytics Process (**Define the Problem**, **Prepare for Modeling**, and **Develop the Model**) and will provide JMP tips or instructions as new tools are introduced. Finally, in Chapter 10, we provide a comprehensive case study that uses the entire Business Analytic Process and introduces other real-life case studies and examples.

References

Chapman, Pete, Julian Clinton, Randy Kerber, Thomas Khabaza, Thomas Reinartz, Colin Shearer, and Rüdiger Wirth. 2000. *CRISP-DM 1.0 Step-by-step data mining guides*. Available at http://ibm.co/1fX7BXN; accessed 09/2014.

Hoerl, R. W., R. D. Snee, and R. R. DeVeaux. 2014. "Applying Statistical Thinking to Big Data Problems." Wiley Interdisciplinary Reviews: *Computational Statistics*, 6(4), 222-232.

INFORMS Certified Analytics Professional Web Page. Available at https://www.informs.org/Certification-Continuing-Ed/Analytics-Certification.

Linoff, Gordon, and M. Michael Berry. 2011. *Data Mining Techniques: For Marketing, Sales, and Customer Relationship Management*, 3rd ed., Wiley. Chapter 1.

SAS Instsitute Inc. 1998. *Data Mining and the Case for Sampling*. Available at http://sceweb.uhcl.edu/boetticher/ML_DataMining/SAS-SEMMA.pdf.

Shearer, C. 2000. "The CRISP-DM model: the new blueprint for data mining." *Journal of Data Warehousing*, 5:13-22.

Part 2

Preparing for Modeling

Chapter 3   Working with Data

In **Part II** we introduce basic navigation and the use of JMP. Then we cover a variety of tools for data visualization, exploration, and basic statistical analysis. Finally, we discuss some common issues with data quality and introduce some tools for data preparation, an essential step before beginning the model building process.

Working with Data

Introduction

One of the most important, and often time-consuming, steps in the Business Analytics Process is preparing data for modeling. This first involves identifying what data are needed to address the business problem, and then either compiling data from available sources, collecting new data, or both. It may require designing and conducting surveys or experiments to collect the needed information, and generally involves process owners and experts from core parts of the business.

As data are compiled, a variety of graphical and statistical tools are used to identify potential gaps (the need for additional or different data), to explore potential data quality issues, and to develop an in-depth understanding of the data and the variables. In this chapter, we'll see how to explore data one variable at a time, two variables at a time, and many variables at a time. We'll introduce data filters and other tools that can be used to dynamically explore subsets of the data and "what if" questions. We'll identify some inherent data quality issues, including missing data, messy data, and data that is in the wrong form or format, and we'll see how to address these issues. We'll also introduce some tools for dealing with high-dimensional data. This refers to data sets that involve many variables, some of which may be related to one another or might even be redundant.

We start with an introduction to JMP. If this is your first exposure to JMP, we suggest that you install JMP and follow along. JMP runs natively on both Macintosh and Windows operating systems (see jmp.com/support for system requirements). There are many wonderful resources to help you get started, so we only highlight the key points here. An in-depth introduction to JMP, *Discovering JMP*, is available online or in the JMP Help files (under **Help > Books**). Additional resources on specific topics, along with short videos, can be found in the JMP Learning Library at jmp.com/learn.

Throughout this book, we'll be using JMP Pro. JMP Pro is the professional version of JMP, which provides advanced tools for modeling and analysis. Note that, for the most part, only Chapters 8 through 10 take advantage of features in JMP Pro. Chapters 3 through 7 primarily use features available in the standard version of JMP.

JMP Basics

Opening JMP and Getting Started

When you first open JMP, you'll see some windows to help you get started.

- **Tip of the Day** gives helpful hints on using JMP. You can close this window, but we recommend that you read through these tips at some point for useful time savers.

- The **JMP Starter** window appears by default on the Mac (click **View > JMP Starter** in Windows to open). For beginning users, this window provides shortcuts for using JMP, including opening files and accessing JMP analyses. Throughout this book, we'll close this window and use options from the JMP menus.

- The **JMP Home** window displays recently used files and open data tables and windows. This window helps you navigate between data tables and analyses. In Windows, closing the home window closes JMP. On the Mac, if you close the home window, JMP will keep running. You can re-launch the home at any time (from **Window > JMP Home**).

JMP Data Tables

Data sets in JMP, called *data tables*, consist of a data grid with data (on the right) and panels (on the left) that define the data and the variables.

An example data table, **Companies.jmp**, is shown in Figure 3.1 (figure borrowed from the one-page guide, **JMP Data Tables**, found at jmp.com/learn). This data table and others we'll use in this book are available in the sample data directory (under **Help > Sample Data Library**).

Figure 3.1:   **Example Data Table in JMP**

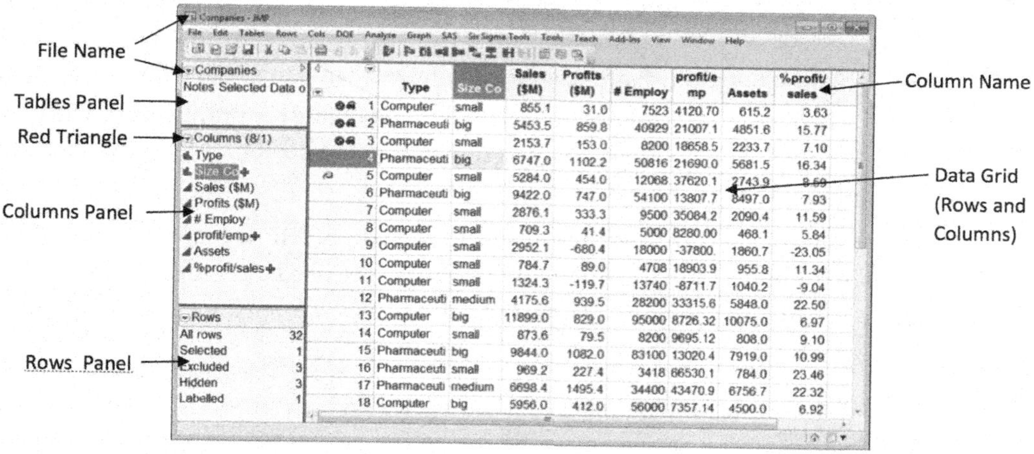

Data in JMP are stored in columns and rows. The column headings (the variable names) are listed across the top, and the row numbers are provided on the side. Each row represents an observation, or a single record.

The panels on the left define our data:

- The **Table Panel** provides the data table name and a list of table properties and saved scripts. (A script is like a macro, which can be saved to the data table. Scripts are written in the JMP Scripting Language, or JSL.)

- The **Columns Panel** tells us about our columns. Here, we can see the number of columns, the number of columns selected, the column names, the modeling type for each column, and column properties that we have specified.

- The **Rows Panel** tells us the number of rows and the number of *selected, excluded, hidden,* and *labeled* rows.

    o   Selected rows are highlighted in the data grid.

    o   Hidden rows, with a mask (🐝) do not appear on graphs.

    o   Excluded rows, with a "don't" sign (⊘) are not included in future analyses.

    o   Labeled rows, with a tag (🏷) are labeled on graphs.

Other elements of the data table, which you'll also see in analyses, are special icons and keystrokes for interacting with JMP:

- *Red triangles*, or hotspots (▾), are used throughout JMP to access other commands. The options provided are context-based and will depend on the location of the red triangle selected.

- *Gray triangles* (▼) are used to minimize or maximize display areas.

- *Right-click* in different regions of the data table (or graphs) for additional options. The options provided depend on the position of your cursor when you right-click.

JMP data tables are stored with a .jmp extension. But, JMP is capable of reading data in a variety of different formats. A few of the commonly used data formats that you can open directly in JMP are provided below. (Search for *Import Your Data* in the **JMP Help** for a complete list.)

- Comma-separated (.csv)

- Microsoft Excel 1997 through 2011 (.xls, .xlsx on Macintosh)

- Microsoft Excel 2007 through 2013 (*.xlsx, *.xlsm on Windows)

- Plain text (.txt)

- SAS versions 7 through 9 on Macintosh (.sas7bdat)

- SAS versions 7 through 9 on Windows (.sas7bdat, .sas7bxat)

- HTML (.htm, .html)

To open an existing JMP data table, select **File > Open**, and navigate to the directory where your data table is stored. Then, select the file and click **Open**. To create a new JMP data table, select **File > New > Data Table** (or **File > New > New Data Table** on the Mac). This will open an empty data table with one column and no rows.

For additional information on JMP data tables, refer to the resources discussed earlier (the book *Discovering JMP* or the JMP Learning Library at jmp.com/learn).

Examining and Understanding Your Data

With the basics of JMP behind us, let's move right into an example.

Historical data were gathered on 5960 bank customers to determine whether a customer is a good or bad credit risk for a home equity loan. Bad risk customers are more likely to default on the loan. The data are stored in the file **Equity.jmp** in the sample data directory (under **Help > Sample Data Library**).

The response (or Y) variable is **BAD**, which is coded as 0 (good risk) or 1 (bad risk). The other variables are:

LOAN: The amount of the loan requested

MORTDUE: How much the customer needs to pay on their mortgage

VALUE: Assessed valuation

REASON: Debt consolidation or home improvement (DebtCon or HomeImp)

JOB: Broad job category

YOJ: Years on the job

DEROG: Number of derogatory reports

DELINQ: The number of delinquent trade lines (or credit accounts)

CLAGE: Age of oldest trade line (oldest credit account)

NINQ: Number of recent credit inquiries

CLNO: Number of trade lines

DEBTINC: Debt to income ratio

Variables in JMP have one of three modeling types: *Continuous*, *Ordinal*, or *Nominal*. In analysis platforms, JMP determines the correct analysis based on the modeling type, so this piece of information is critical.

- *Continuous* variables, like **LOAN** and **MORTDUE**, have numeric values (e.g., 2, 5, 3.35, 159.667,…). These variables are denoted with a blue triangle (◀) in the data table columns panel and in analysis dialogs. Continuous variables must be in numeric format and can have no special symbols or text.

- *Nominal* variables, like **BAD**, **REASON** and **JOB**, can have either numeric or character values, and represent unordered categories or labels (e.g., the names of states, colors of M&Ms, machine numbers, and so on). Nominal variables are denoted with red bars (🟥).

- *Ordinal* variables, which are denoted with green bars (🟩) can also have either numeric or character values. Ordinal variables represent ordered categories (e.g., small, medium, and large; 1-9 severity rating scales, and so on). This data set has no ordinal variables.

A first step in any analysis is to ensure that your variables have the correct modeling type. To change the modeling type, click on the icon next to the variable in the data table or in any analysis dialog window. JMP also allows for a number of data types in addition to modeling types. These data types can be changed using the **Column Info** window (right-click on a column name and select **Column Info**). Note that there are a number of column properties that can be specified from this window, including notes and formulas.

Exploring Data One Variable at a Time

We start by looking at the variables one at a time. We use summary statistics and graphical summaries to get familiar with our data and at the same time identify any potential data quality issues.

For continuous variables, we're interested in summary statistics, such as the mean (the average), the standard deviation (a measure of the spread), minimum and maximum values, and the number of missing observations. We're also interested in the shape of the distribution. Is the distribution more or less symmetric? Is it skewed? Are there clusters of data or severe outliers? Are there values that aren't physically possible?

For categorical data (nominal or ordinal modeling types), we're interested in the number of categories (or levels), the number of observations in each category, and the number of missing observations.

Two key tools in JMP for exploring variables one at a time are **Columns Viewer** and **Distribution**. The **Columns Viewer** provides numeric summaries of our data, as shown in Figure 3.2, and is particularly useful with data sets with many variables. (Go to **Cols > Columns Viewer**. Select all of the variables from the **Select Columns** list, click **Show Quartiles**, and then click **Show Summary**.)

For the three nominal variables, **BAD, REASON**, and **JOB**, we see the number of observations (**N**), the number of missing values (**N Missing**), and the number of categories (**N Categories**). For the continuous variables, we also see **N** and **N Missing**, plus a number of summary statistics. Of potential concern is the number of missing records, particularly for **DEBTINC**.

The **Min** and **Max** provide the range of values for the continuous variables. Many of these variables have a minimum value of 0, and some of these have medians and quartiles that are also 0. This may or may not be an issue, but it should be investigated. The mean is the average value, while the median is the middle value (the 50th percentile). Big differences between these two values would be an indication of potential skewness.

Figure 3.2:   Columns Viewer, Equity Data

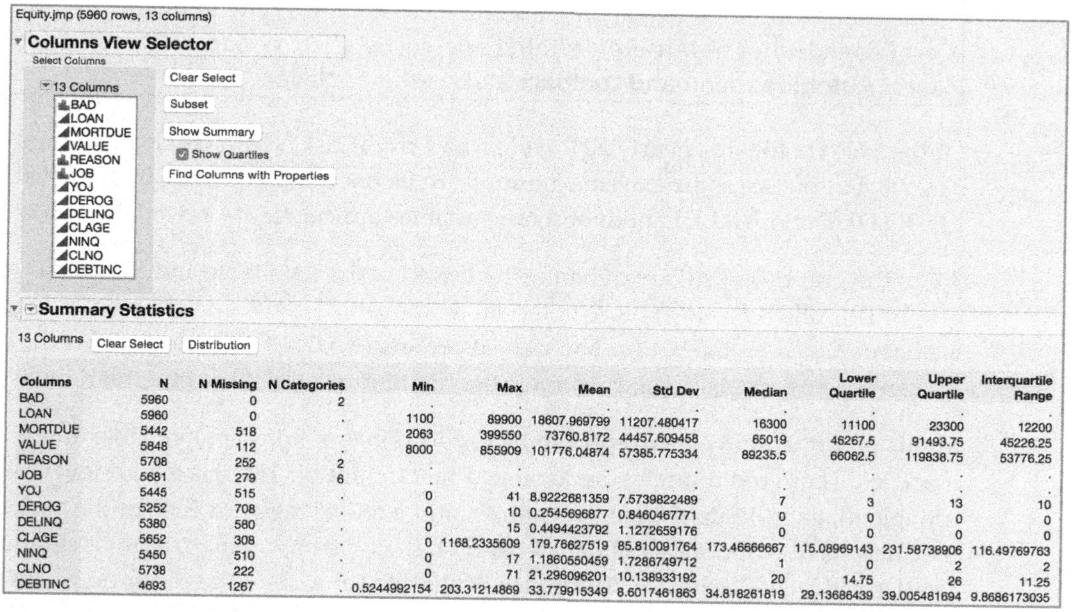

Equity.jmp (5960 rows, 13 columns)

▼ Columns View Selector

Select Columns

▼ 13 Columns   Clear Select

    ▲BAD   Subset
    ◢LOAN
    ◢MORTDUE   Show Summary
    ◢VALUE
    ▲REASON   ☑ Show Quartiles
    ▲JOB   Find Columns with Properties
    ◢YOJ
    ◢DEROG
    ◢DELINQ
    ◢CLAGE
    ◢NINQ
    ◢CLNO
    ◢DEBTINC

▼ ▼ Summary Statistics

13 Columns   Clear Select   Distribution

Columns	N	N Missing	N Categories	Min	Max	Mean	Std Dev	Median	Lower Quartile	Upper Quartile	Interquartile Range
BAD	5960	0	2	.	.	.	.	.	.	.	.
LOAN	5960	0	.	1100	89900	18607.969799	11207.480417	16300	11100	23300	12200
MORTDUE	5442	518	.	2063	399550	73760.8172	44457.609458	65019	46267.5	91493.75	45226.25
VALUE	5848	112	.	8000	855909	101776.04874	57385.775334	89235.5	66062.5	119838.75	53776.25
REASON	5708	252	2	.	.	.	.	.	.	.	.
JOB	5681	279	6	.	.	.	.	.	.	.	.
YOJ	5445	515	.	0	41	8.9222681359	7.5739822489	7	3	13	10
DEROG	5252	708	.	0	10	0.2545696877	0.8460467771	0	0	0	0
DELINQ	5380	580	.	0	15	0.4494423792	1.1272659176	0	0	0	0
CLAGE	5652	308	.	0	1168.2335609	179.76627519	85.810091764	173.46666667	115.08969143	231.58738906	116.49769763
NINQ	5450	510	.	0	17	1.1860550459	1.7286749712	1	0	2	2
CLNO	5738	222	.	0	71	21.296096201	10.138933192	20	14.75	26	11.25
DEBTINC	4693	1267	.	0.5244992154	203.31214869	33.779915349	8.6017461863	34.818261819	29.13686439	39.005481694	9.8686173035

The **Distribution** button (above the statistics in Figure 3.2) launches the **Distribution** platform for selected variables. This platform is also available from the **Analyze** menu.

In Figure 3.3, we see distributions for the first five variables. (In **Analyze > Distribution**, select the variables, click **Y, Columns**, and then click **OK**.) For categorical variables, JMP produces bar charts and frequency distributions. For continuous variables, JMP produces histograms, box plots, and summary statistics.

By default, the output is displayed vertically. This allows us to see distributions of several variables at one time. However, you can easily convert this to a horizontal view (click the top red triangle and select **Stack**). To produce horizontal histograms from **Distribution** in the future, you can set a preference.

To set preferences in JMP, go to **File > Preferences** (or **JMP > Preferences** on the Mac). From here, you can change font sizes and types, default output, graph formats, and other customizations. To change the default layout for **Distribution** reports to horizontal, select **Platforms** from the **Preference Group**, select **Distribution** from the list, and check the **Stack** and **Horizontal Layout** boxes. Finally, click **Apply** to accept and then click **OK** to close.

Note to Windows Users: By default, your menus and toolbars will be hidden in analysis windows. To display menus and toolbars in every window (this is recommended for new users), within preferences, select **Windows Specific**. Then select **Autohide menu and toolbars** and set this to **Never**.

Of the 5960 customers, nearly 20% were a bad credit risk. The mean loan amount is $18,608, but we see some extreme amounts. In fact, we also see some extreme values for **MORTDUE** and **VALUE**. Both of these variables appear right-skewed.

All of the graphs in JMP are dynamically linked to the data table and to every other graph. This allows us to explore potential relationships between variables. In Figure 3.3, we have clicked on the bar for bad risk customers (**BAD = 1**). This highlights the bar, and we can see how the bad risk customers are distributed across the other variables.

In this report, and in every report in JMP, additional options are available under the red triangles. These red triangles are arranged hierarchically. There is a high-level red triangle that applies to the entire window, and a red triangle for each of the variables in the analysis (the options available depend on the type of variable). In this window, there are also red triangles next to **Summary Statistics** for each of the continuous variables.

Figure 3.3:   Distribution Output, First Five Variables

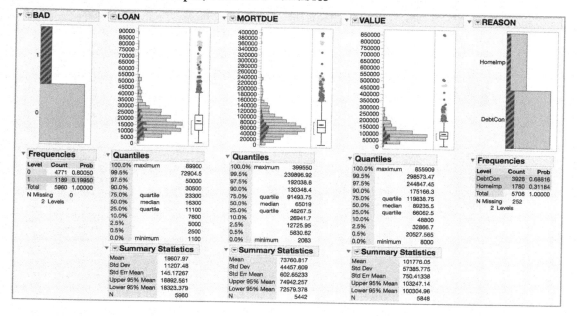

In Figure 3.4, we see **Distribution** output for the next five variables in the data set. Recall from Figure 3.2 that some of the continuous variables, like **DEROG** and **DELINQ**, have minimum values of zero. For these variables, we can clearly see the distributions using the histograms and quantiles. Most of the values are, in fact, zero. In **CLAGE**, we see another potential problem: 2 customers with loan ages near 1200 days (these can be seen as points in the box plot).

So far, we've identified four potential data quality issues:

- Missing values for many variables
- Skewed distributions, and a long right-tail in the distributions of some variables
- Messy data, including continuous variables with many zeros
- Outliers in **CLAGE**

Figure 3.4: Distribution Output, Next Five Variables

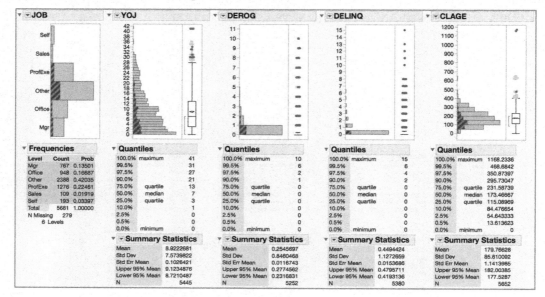

Saving Your Work

Before proceeding, it's a good idea to save our work. We can do this in a number of ways. Here, we briefly describe two approaches for saving your work:

1. *Copy and Paste Output into Another Program*: Use the selection tool in the toolbar (✛) to click and select content that you'd like to copy (selected content is highlighted). Open the program where you'd like to paste the content, and select Paste. For example, to paste as an image in MS Office programs, select **Paste > Paste Special** and from the list select the desired format (for example, **Picture (Enhanced Metafile)**). If you're using a Mac, select **Paste**, or use **Paste Special > PDF** for high-quality graphics.

2. *Save the Script to the Data Table*: Saving the steps taken to produce a report as a script enables you to re-create the report at any time. From any JMP output window, click the top red triangle and select **Script > Save Script to Data Table**. The saved script will appear in the table panel of the data table (in the top left corner). To run the script, click on the red triangle and select **Run Script**.

Hint: Save the data table to save the script and any other changes you've made to the data table. Note that scripts will perform the same analysis steps on the current data, so if you change your data values, or hide and/or exclude values, the new analysis may provide different results.

Exploring Data Two Variables at a Time

As we saw in the previous section, dynamic linking allows us to start to get a feel for potential bivariate relationships between our variables. Three additional platforms for exploring variables two at a time are **Tabulate**, **Graph Builder**, and **Fit Y by X**.

Tabulate, under the **Analyze** menu, is a platform for dynamically summarizing (tabulating) data and constructing tables of descriptive statistics. The initial **Tabulate** window is shown in Figure 3.5. To summarize data, drag variables from the columns list to the drop zones for columns and rows, then drag statistics of interest into the results zone to add new statistics. Then, click **Done** to close the control panel and produce the final table.

In Figure 3.6, we see percentages for **BAD** for the two reasons and six job categories. To re-create this tabulation:

1.   Drag **BAD** to **Drop Zone for Columns**.

2.   Drag **REASON** to **Drop Zone for Rows**.

3.   Drag **JOB** just below **REASON**.

4.   Drag **Row %** and **N** to the results area.

There are more loans for customers consolidating debt than home improvement, and two job categories, **Sales** and **Self**, appear to have higher risk of bad loans than the other categories. However, there are far fewer customers in these two job categories.

Figure 3.5: Initial Tabulate Window

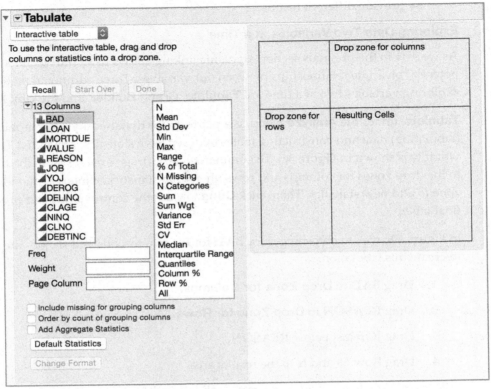

Figure 3.6: Tabular Summary of BAD versus REASON and JOB

	BAD			
	0		1	
REASON	N	Row %	N	Row %
DebtCon	3183	81.03%	745	18.97%
HomeImp	1384	77.75%	396	22.25%
JOB				
Mgr	588	76.66%	179	23.34%
Office	823	86.81%	125	13.19%
Other	1834	76.80%	554	23.20%
ProfExe	1064	83.39%	212	16.61%
Sales	71	65.14%	38	34.86%
Self	135	69.95%	58	30.05%

The means of the continuous variables for the two categories of **BAD** are shown in Figure 3.7. There appear to be some potential differences. For example, the loan amounts are higher on average for low-risk loans, while the debt-to-income ratio is higher for high-risk loans. These bivariate summaries highlight some potentially important variables.

Figure 3.7:   Means of Continuous Variables for Levels of BAD

▼ ⊡ **Tabulate**			
		BAD	
		0	**1**
LOAN	Mean	19028.11	16922.12
MORTDUE	Mean	74829.25	69460.45
YOJ	Mean	9.15	8.03
DEROG	Mean	0.13	0.71
DELINQ	Mean	0.25	1.23
CLAGE	Mean	187.00	150.19
NINQ	Mean	1.03	1.78
CLNO	Mean	21.32	21.21
DEBTINC	Mean	33.25	39.39

The **Graph Builder** is the first option under the **Graph** menu. This is a flexible and highly interactive graphing and exploratory platform. The initial window is shown in Figure 3.8. Like **Tabulate**, the **Graph Builder** uses drop zones for specifying the variables to display. The primary zones are **Y** and **X**, but there are other zones along the top and right that allow you to add additional information to the graph. The icons across the top can be used to change the graph type displayed, and additional options for the selected graph element are provided on the bottom left, below the variable list. The icons that are available depend on the modeling types of the variables selected.

In Figure 3.9, we see box plots for **DEBTINC** and **LOAN**. To re-create this graph:

1. Drag **BAD** to **Y**.

2. Drag **DEBTINC** to **X**.

3. Drag **LOAN** to the right of **DEBTINC** on the **X** zone.

4. Click **Done** to close the control panel.

We can see that more of the high risk customers had a high debt-to-income ratio, and more of the high loan amounts were for low risk customers.

Figure 3.8:   Initial Graph Builder Window

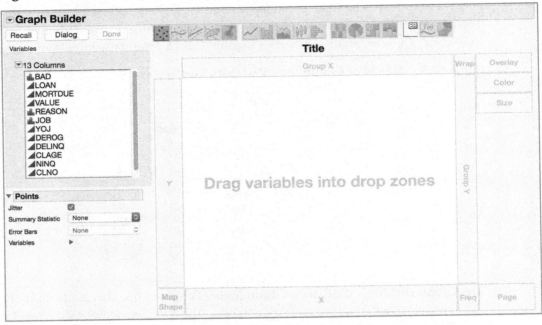

Figure 3.9:   Box Plots for DEBTINC and LOAN versus BAD

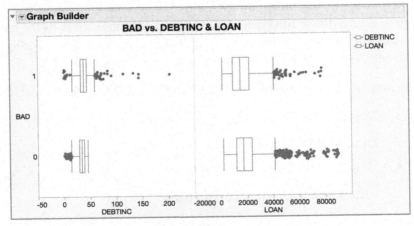

In Figure 3.10, we see a plot of **BAD** versus **JOB** (drag **BAD** to **Y**, **JOB** to **X**, select the **Mosaic** graph element from the icons at the top, and then click **Done**). This plot, called a mosaic plot, is an efficient way to graphically explore the relationship between two categorical variables. The widths of the bars across the bottom show the relative frequencies of the job categories. The bars are broken down, in this case, by **BAD = 0** (the bottom area of the bars) and **BAD = 1** (the top area). From this, we can see that there are very few customers in sales or customers who are self-employed. We also see that the undefined **Other** category has the most customers, and the lowest risk customers are office workers and professionals or executives. (Hint: Double-click on the *x*-axis to change the axis settings. Here, we've changed the **Label Orientation** to **Angled** to display all of the labels.)

Figure 3.10:   Mosaic Plot of JOB versus BAD

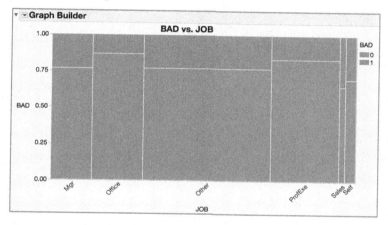

With **Graph Builder**, we have explored the graphical relationship between some of our variables and our nominal response. Additional graphical options are available, including line plots, bar charts, histograms, and background maps. And, several variables can be explored at one time.

To formally analyze the relationship between two variables, we use the **Fit Y by X** platform from the **Analyze** menu. Like most platforms in JMP, **Fit Y by X** is contextual, producing analyses based on the modeling types of the variables selected.

The **Fit Y by X** dialog window is shown in Figure 3.11. The icons on the bottom left indicate what type of analysis JMP will provide based on the modeling types of the response and the factor. In this example, we have selected a nominal response (**BAD**), so the two available analyses are logistic regression and contingency analysis. We have selected one continuous factor (**LOAN**) and one nominal factor (**JOB**). JMP will produce both logistic regression (for the continuous factor) and contingency (for the nominal factor).

Figure 3.11:   Fit Y by X Dialog Window

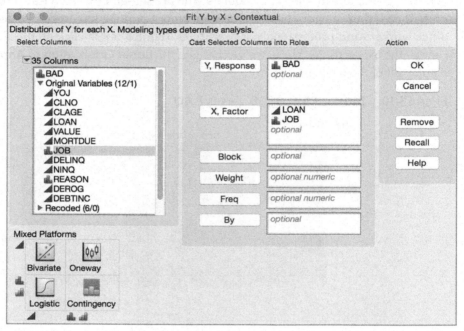

In Figure 3.12, we see partial results for both analyses. The mosaic plot is the same plot that we created using the **Graph Builder**, but summary information and statistical output are also provided, along with many options under the red triangle.

The logistic plot shows what happens to the probability that a customer will be low risk (**BAD** = 0) as the loan amount increases. Since the slope of the curve is positive, the probability increases (meaning that the probability that the customer will be a bad risk (**BAD** = 1) decreases. We'll provide background technical details on logistic regression in Chapter 5.

Figure 3.12: Fit Y by X, BAD versus LOAN and JOB

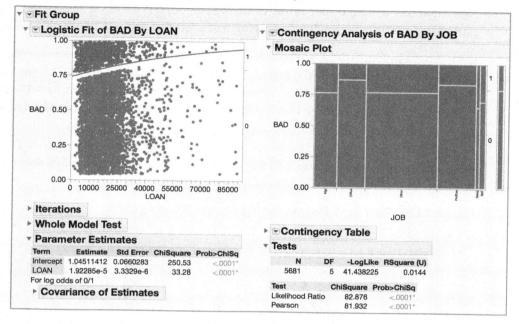

If our response is continuous, our two analysis options from **Fit Y by X** are *Bivariate* and *Oneway*. Bivariate produces a scatterplot, and provides options for correlation and regression. Oneway provides graphical options like box plots and many important analysis options, such as two-sample t-Tests and analysis of variance (ANOVA).

For illustration, we briefly leave the equity data and use an example with a continuous response (we'll return to **Equity.jmp** in a moment). The **Boston Housing.jmp** data, from the **Sample Data** directory, is popular for teaching regression and other modeling techniques. The response of interest is the continuous variable **mvalue**, which is the median home value (in $1000) for towns in the Boston area in the 1970s. There are several potential predictors, which we'll describe in a future chapter. Two predictors of interest are the number of rooms (**rooms**) and a dummy variable indicating whether the town tracks the Charles River (**chas=1** indicates the town tracks the river).

On the left in Figure 3.13, we see partial results from the bivariate analysis, where **mvalue** is the response and **rooms** is the factor. Here, we've fit a regression line (select **Fit Line** from the red triangle). On average, as the number of rooms increases, the median value of homes also increases. But this plot provides additional information.

For example, we see a number of towns with median home values of 50 ($50,000). It is likely that this was an upper cutoff value of some sort. We'll cover regression and, in particular, multiple regression (involving many predictors), in Chapter 4.

On the right in Figure 3.13, we see partial results from the oneway analysis with **mvalue** as the response and **chas** as the factor. We have selected **Quantiles** and **t Test** from the top red triangle (and have selected **Points Jittered** from the **red triangle > Display Options**). The *p*-value for the t-Test indicates that there is a significant difference in median home value for **chas = 0** versus **chas = 1** (the *p*-value for the two-tailed test is reported under Prob > |t|). This graph also provides additional information. For example, the number of points and the width of the *x*-axis labels indicate that there are far more towns that do not track the Charles River than towns that do.

Figure 3.13:   Fit Y by X Output for the Boston Housing Data

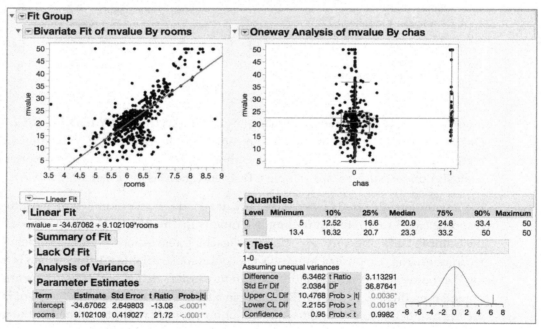

For both types of analyses, many more options are available under the red triangle for that analysis. We encourage you to explore these options on your own. For background information behind the statistical methods available, see the **JMP Help** (**Books > Basic Analysis**) or consult an introductory statistics text. For a listing of statistical methods that are available in JMP, see the **Statistics Index** under the **Help** menu.

Exploring Data Three or More Variables at a Time

Tabulate and **Graph Builder**, introduced in the previous section, can also be used to explore more than two variables at a time. To illustrate, we return to the **Equity.jmp** data. In Figure 3.14 (left), we see a tabular summary of **BAD** by **REASON** and **JOB** within **REASON**, along with average loan amounts. On the right in Figure 3.14, we have a graphical view of the same four variables, where the bar heights represent the number of customers and the bar colors represent the average loan amount.

Figure 3.14: Tabulate and Graph Builder with Multiple Variables

Tabulate

		BAD		
		0	1	LOAN
REASON	JOB	Row %	Row %	Mean
DebtCon	Mgr	0.80	0.20	20799.30
	Office	0.85	0.15	19056.29
	Other	0.79	0.21	20266.08
	ProfExe	0.85	0.15	20151.83
	Sales	0.65	0.35	16102.06
	Self	0.56	0.44	29389.04
HomeImp	Mgr	0.68	0.32	14194.83
	Office	0.89	0.11	16313.95
	Other	0.71	0.29	13175.42
	ProfExe	0.82	0.18	16594.07
	Sales	0.67	0.33	5308.33
	Self	0.82	0.18	27649.57

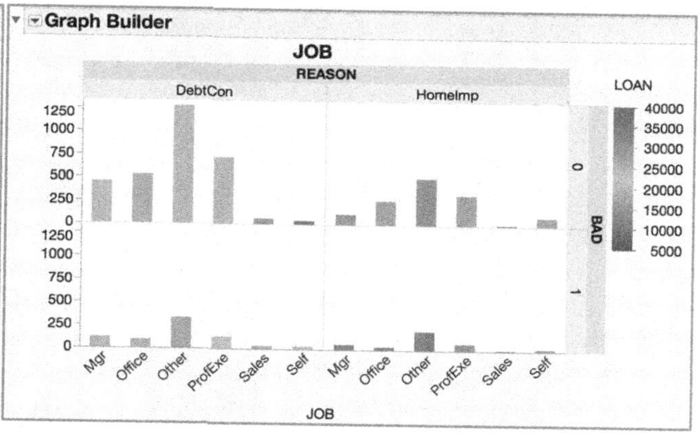

A popular display of potential relationships between continuous variables is a scatterplot matrix (from **Analyze > Multivariate Methods > Multivariate**, select continuous variables as **Y, Columns**, and click **OK**). A scatterplot matrix for the first five continuous variables is shown in Figure 3.15 (bottom).

A table of correlations for the variables is also provided (top, in Figure 3.15). The points are colored by **BAD** = **0** (red) and **BAD** = **1** (blue). Each box graphically shows the correlation, or the linear association, between two variables. The first row contains scatterplots between **LOAN** and each of the other variables, the second row is between **MORTDUE** and the other variables, and so on. Likewise, the first row in the table of correlations shows the correlation between **LOAN** and each of the other variables. A density ellipse is drawn in each scatterplot to provide an overall view of the relationship or association between each pair of variables. Ellipses for strongly correlated variables are narrow, while ellipses for variables that are weakly correlated are more circular.

Figure 3.15: Equity Scatterplot Matrix

Multivariate

Correlations

	LOAN	MORTDUE	VALUE	YOJ	DEROG
LOAN	1.0000	0.2286	0.3354	0.1057	-0.0013
MORTDUE	0.2286	1.0000	0.8757	-0.0885	-0.0495
VALUE	0.3354	0.8757	1.0000	0.0078	-0.0487
YOJ	0.1057	-0.0885	0.0078	1.0000	-0.0659
DEROG	-0.0013	-0.0495	-0.0487	-0.0659	1.0000

There are 1395 missing values. The correlations are estimated by Pairwise method.

Scatterplot Matrix

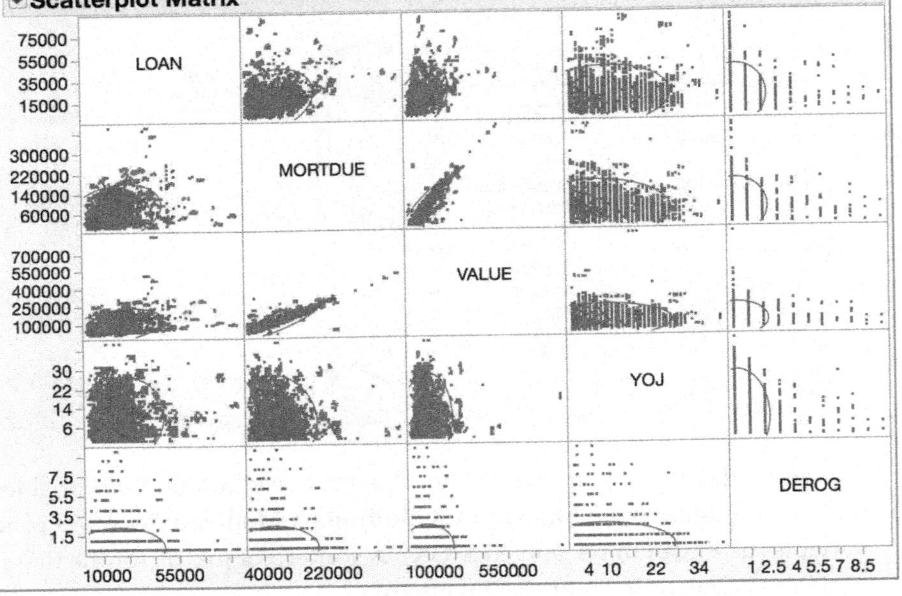

In the scatterplot, we see that there is a strong relationship between **MORTDUE** and **VALUE** and relatively weak relationships between **MORTDUE** and the other variables. This type of chart is also good for finding potential outliers in the data. For instance, there are some high values for both **MORTDUE** and **VALUE**, and some values that are high for **LOAN** and low for **MORTDUE**.

Figure 3.15 displays points for both low-risk (**BAD** = **0**) and high-risk (**BAD** = **1**) customers. To display the relationship between the variables for only high-risk customers, we use the **Local Data Filter** (under the red triangle next to **Multivariate > Script** or via the **Local Data Filter** icon on your tool bar). The **Local Data Filter** allows you to update an open graph or analysis, by showing or including only rows associated with specified values of a selected variable. This allows you to dynamically explore results based on values of other variables.

In Figure 3.16, we use the **Local Data Filter** to show and include only the observations of the variables when **BAD** = 1 (select **BAD** as the filter column, click **Add**, and then select **BAD** = 1).

Figure 3.16:   Equity Scatterplot Matrix with Local Data Filter

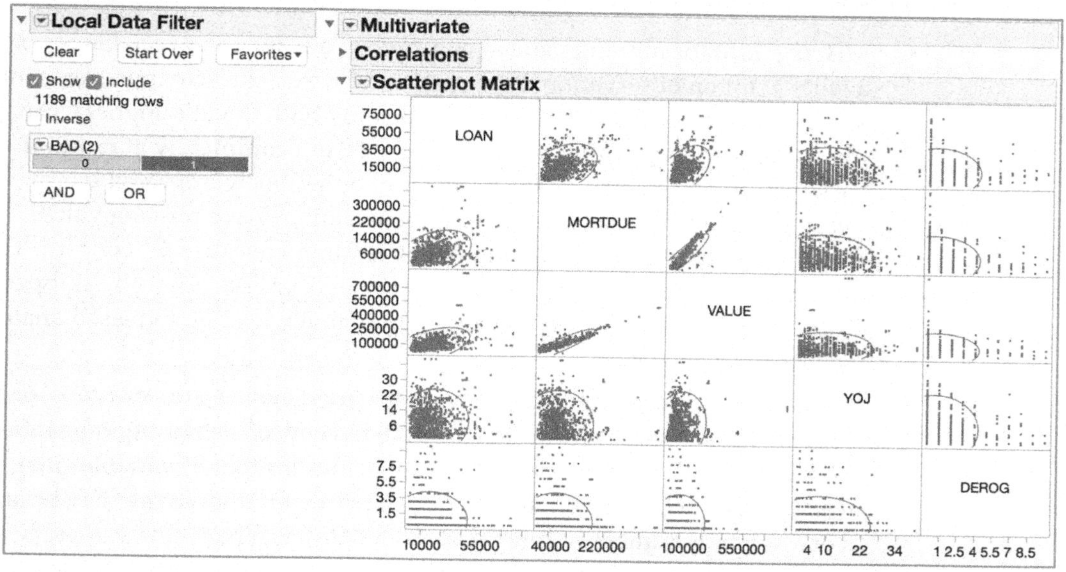

Note that the **Local Data Filter** interacts only with the graph or analysis that is displayed in the report window. A global version of the **Data Filter**, which applies to the data table and all open graphs and analyses, is available under the **Rows** menu.

Preparing Data for Modeling

In any study, it's important to know how the data were collected and exactly what they represent. For example, were the data recorded in a meaningful way? Are the data accurate? Do the measurements correctly capture or reflect the characteristic of interest? Do the data reflect the current process or system?

In addition to accuracy-related issues, there are a variety of other potential problems that may need to be addressed before we begin the modeling process. These issues can stem from a variety of causes, such as manual data entry, uncalibrated equipment, inconsistent methods or procedures, conversion of data from one system to another, and system consolidation.

Common Problems with Data

We have uncovered some potential problems in our exploration of the equity loan data. Some of the more common data quality problems, along with issues that need to be addressed during data preparation, are briefly described below. Note that we discuss how to address some of these issues in the next section.

- *Missing data* (or missing values) occur when a value for a variable is not available for an observation. Understanding why values are missing is important and can lead to different strategies for dealing with the missingness. Data may be **missing at random** within a variable, or **missing completely at random** across variables. In these cases, analyses may not be seriously impacted by the missingness unless a large number of observations have missing values. However, values that are missing **not at random** are cause for concern. This occurs, for example, when a question on a survey is purposefully skipped or when sensitive information is omitted, potentially leading to biased analyses or incorrect conclusions.

- *Dirty* or *messy data* are data that are inaccurate, have errors or typos, or are not complete. Data may be incorrectly coded, have inconsistent capitalization, abbreviations and spacing. Records may be duplicated, or variables may be redundant or highly correlated. For categorical variables, there may be an overwhelming number of categories, some of which have few values. Continuous data may be highly skewed or multi-modal, and can have extreme observations or clumps of observations.

- *Incomplete data* can relate to missing variables or not having enough data to perform an analysis. If critical variables are missing, the predictive model will most likely not perform well. If there aren't enough observations, it may not be possible to estimate important model parameters and the model predictions may not be very precise.

- A flip side of this is *extremely large data sets* in terms of **many variables** or **many observations**. While this isn't a shortcoming of the data or a data quality issue per se, it can pose a challenge for modeling. Having too many variables can be problematic when the variables are correlated with one another, are redundant, or don't provide any useful information about the response. However, variable reduction methods can be applied, and "modern" modeling techniques are effective in dealing with a large number of observations and predictors.

- *Incorrectly formatted* data are data in the wrong form or format for analysis. This can apply to the data table as a whole, or to the formatting of variables in the data table. For example, data might be stored in separate columns while an analysis requires data stacked in one column. Or individual variables in JMP might have the incorrect modeling type. Since the modeling type drives the analysis in JMP, having an incorrect modeling type may lead to the wrong analysis. Another issue relates to dates and times: Columns containing date or time measurements need to be formatted as date or time variables (in **Column Info**) in order to perform calculations, such as elapsed time.

This is, by no means, a comprehensive list. It's based primarily on the experience of the authors. Working through data quality issues and preparing data for modeling can be very time consuming, and the importance of this effort cannot be overestimated (recall the expression "garbage in garbage out").

In this section, we introduce tools for addressing some of these issues in JMP, specifically:

- Working with missing values
- Transforming or deriving new variables
- Binning continuous variables
- Reducing the number of variables (dimension reduction)
- Using options from the **Tables** menu to restructure data tables

Working with Missing Values

The **Recode** column utility can be used to fix issues with capitalization, combine categories, and perform other general housekeeping with categorical variables. It can also be used to create a **Missing** category for categorical variables, as shown in Figure 3.17.

The two categorical variables, **JOB** and **REASON**, along with many of the continuous variables, are missing values. To create a **Missing** category for **JOB**, select the variable in the data table, and then select **Cols > Utilities > Recode**. The **Old Value** in the first row is blank, indicating that the values are missing. Type "Missing" as the **New Value** in the first row, and then select an option under **Done** to save your work. (Note that this variable has a **List Check** column property. This will prevent recoding. If you're following along, before recoding, go to **Column Info** for **JOB**, select the column property, and click **Remove** to remove the **List Check** column property.)

Figure 3.17:   Recoding Missing Values in JOB

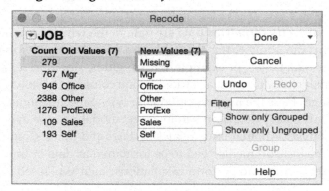

To save this recoding as a formula in a new column, select **Done > Formula Column**.
The original column name in this example is **JOB**, so this creates a new column at the end
of the data table named **JOB 2** with a stored formula (we have renamed the column **JOB
Recode**). Recode has used a built in function, **Conditional > Match**, to create the formula
(shown in Figure 3.18).

Figure 3.18:   Formula for Recoding JOB

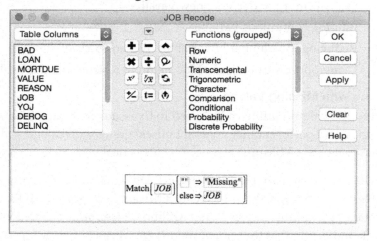

When a continuous variable has missing values, imputation is often used to replace the
missing values with substituted values. Imputation in JMP Pro is available from the
Explore Missing Values option under **Cols > Modeling Utilities**, and from many
modeling platforms.

Informative Missing is another approach for handling missing values. How the missing values are handled depends on whether the variables are continuous or categorical. For continuous variables, the model will include the original variable with the mean value imputed for missing values and a second variable indicating whether the value is missing or not. For categorical variables, an additional level will be included that indicates that the value for the variable is missing. You could create these columns yourself using recoding and the formula editor, but JMP platforms with the option for informative missing do this automatically.

Additional details on informative missing are provided in the Chapter 6, "Decision Trees." For more on handling missing values in JMP and JMP Pro, search for **Explore Missing** or **Informative Missing** in the **JMP Help**.

Transforming (or Deriving New) Variables

Some of our variables, such as **LOAN**, are highly skewed (see Figures 3.3 and 3.4). Applying a log (or other) transformation can make distributions appear less skewed. While it isn't necessary for predictors to be normally distributed, transformations can make it easier to see patterns in the data and can help us make sure our assumptions for using particular statistical methods are met (note that we'll talk more about these assumptions in the modeling chapters).

If you know which transformation you'd like to apply, variables can be transformed directly from the data table, as shown in Figure 3.19. Simply right-click on the column, and select **New Formula Column > Transform > Log** (or another transformation). A new column will be created with the formula for the transformation.

In many cases, some exploratory work can help identify the best transformation. From any graph or analysis launch window, you can right-click on a variable and select a transformation. This creates a temporary (or *virtual*) variable, which can be graphed or included in an analysis (as shown in Figure 3.20). To save this new derived variable to the data table, right-click on the virtual column and select **Add to Data Table**.

Figure 3.19:   Transforming Loan from the Data Table

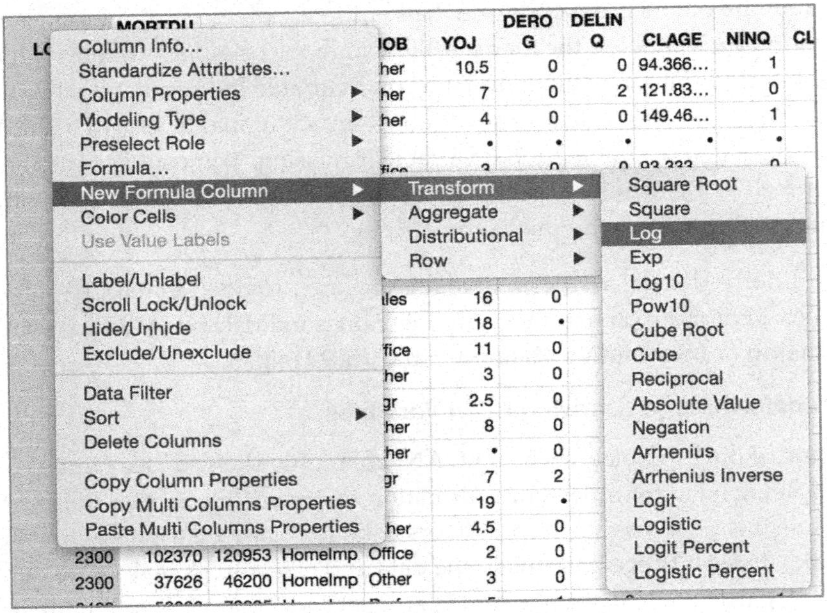

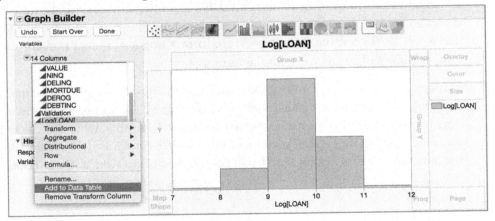

Figure 3.20:   Transforming Loan from the Graph Builder

Shortcuts to other formula editor functions can also be created in this way. For example, if a variable is stored in JMP with a date format, convenient date and time transformations are available.

For more information on transformations to apply in particular situations, and reasons why transformations can be beneficial, see Kutner et al. (2004) or Sharpe et al. (2014).

Binning Continuous Data

Binning is sometimes used when dealing with a continuous variable that has messy or unruly data. In the equity data, for example, **DEROG** and **DELINQ** are coded as continuous data, but a majority of the values are zero (see Figure 3.2). In addition, both of these variables have missing values. To address these issues, we can create a formula to bin the values into categories, with an extra category to include missing values. For example, since a majority of the values are zero and integer values (counts) are used, we can convert this variable into a categorical variable with three categories: **None**, **1 or More**, and **Missing**.

To do this, we add a new column to the data table and use the **Formula Editor** to create a formula. We use two function groups to create the formula: **Conditional** and **Comparison**. The carat (^) on the keyboard is used to add additional arguments to the formula. The formula for binning **DEROG** into three categories is shown in Figure 3.21.

Figure 3.21:   Binning DEROG into Three Categories

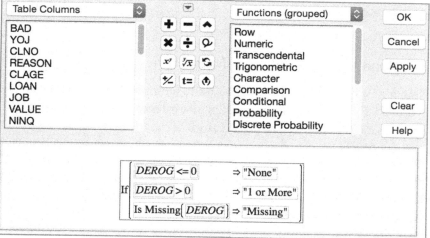

To create this formula:

1. Select **Conditional > If** from the **Functions** list.

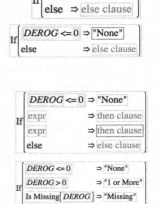

2. Select **DEROG** from **Table Columns**. From the **Functions** list select **Comparison**, **a <= b**, and type a 0. Click in the **then clause** and type "None", and click **Return** (or Tab) to accept.

3. With the argument for "None" still highlighted, click the carat (**^**) symbol on your keyboard four times to insert two new rows.

4. Click on the first new expr argument (in the second row), and select **DEROG**. From the **Functions** list, select **Comparison**, **a > b**, and then type 0. In the **then clause**, type "1 or More".

5. In the second new expr argument (in the third row) select **DEROG**, and then, from the **Functions** list, select **Comparison**, **Is Missing**. In the **then clause**, type "Missing".

6. Delete the **else clause**.

Click **Apply** to view the results in the data table, and click **OK** to accept.

Note that there can be some downsides to binning data, and this technique should not be overused. Binning of continuous data into categories results in loss of information, since we're no longer using the original values.

Using binned data versus the original continuous variables changes the way the model is built. Take, for example, a regression model with one continuous predictor and one five-level categorical predictor. For the continuous predictor, only one model effect is estimated. For categorical predictors, with k levels, k-1 model effects must be estimated. So, in this example, four model effects are estimated for the categorical predictor. This can be advantageous when dealing with extremely large data sets. (And some statistical packages will automatically bin all continuous variables.) However, it can become problematic when building models with small data sets or with many categorical predictors. In this case, models can quickly run out of information, or *degrees of freedom*, to estimate model effects.

For details on creating formulas, search for **Formula Editor** or **Virtual Columns** in the JMP Help or in the book **Using JMP** (under **Help > Books**).

Dimension Reduction

The dimension of a data set is the number of variables. High-dimensional data sets may include variables that are highly correlated or completely redundant with one another, or variables that do not help explain or predict the response of interest. This can lead to a number of problems, including computational issues, models that are overly complex, added expense, and models that don't perform well in practice (see Shmueli et al., 2010)

A first step in reducing the number of predictors is to use subject matter knowledge to understand what the various predictors represent and why they might be important when predicting the response. Some variables may be known to be more relevant than others. There may also be practical or financial reasons for including some predictors over others. Another consideration is the quality of the data. Redundant or correlated predictors that are missing many values, are inaccurate or have other issues are candidates for removal.

It is also helpful to use graphical displays of the predictors and the response. For example, the scatterplot in Figure 3.15 shows that the predictors **VALUE** and **MORTDUE** are highly correlated with one another. Including both of these variables may lead to issues when building models.

A statistical tool for dimension reduction, which is particularly useful when predictors are highly correlated, is *Principal Components Analysis* (or *PCA*). PCA converts the original variables to a few variables, or principal components, which are uncorrelated. These new variables are a weighted combination of the original variables, which retain the information content of the original predictors. This allows us to reduce the total number of variables without losing much explanatory information. The topic of principal components analysis, and dimension reduction techniques in general, deserves more attention than we can provide here. For a detailed discussion, see Shmueli, 2010.

In JMP, **Principal Components Analysis** is available from the **Analyze > Multivariate Methods** menu. We omit discussion of the analysis and introduce a useful variable reduction technique available from the PCA platform in *JMP Pro*, **Variable Clustering**. This option provides groupings of similar variables, identifies the most representative variable in each grouping, and calculates the total variation in the predictors explained by these most representative variables (SAS, 2015).

For an example, we conduct a principal component analysis using 11 continuous predictors in the **BostonHousing.jmp** data set (we omit the variable **b**). **Cluster Variables** is an option under the top red triangle. In Figure 3.22, we see that three variables, **distance**, **radial**, and **rooms** are the most representative variables in each of three groupings of variables. These three variables explain 72.8% of the total variation in the predictors. This information, along with subject matter knowledge, can lead to reduction in the total number of variables considered when building models.

Figure 3.22:   Boston Housing Data Cluster Variables

	▼ ▾ **Principal Components: on Correlations**						

	▶ **Summary Plots**

	▼ ▾ **Variable Clustering**

▼ Cluster Summary

Cluster	Number of Members	Most Representative Variable	Cluster Proportion of Variation Explained	Total Proportion of Variation Explained	.2 .4 .6 .8
1	5	distance	0.735	0.334	
2	3	radial	0.808	0.22	
3	3	rooms	0.636	0.173	

Proportion of variation explained by clustering: 0.728

Note that PCA is used when the correlated predictors are continuous. When the predictors are categorical, multiple correspondence analysis can be used (under **Analyze > Consumer Research > Multiple Correspondence Analysis**).

Restructuring Your Data

Finally, you may find that your data are not in the correct format for analysis. For example, most analyses in JMP require data to be in stacked form, where columns represent measurements and rows represent observations. Or critical pieces of information may reside within different data tables. In JMP, operations are performed on the active data table, so all variables and observations need to be stored within the same data table.

A variety of options for managing or rearranging data are available under the **Tables** menu. For example, **Stack** can be used to stack data stored in separate columns in a single column (with the labels stored in a second column), and **Join** can be used to merge data from the two tables into a new table. For more information on these and other tools in the **Tables** menu, search for **Reshaping Data** in the **JMP Help** or in the book **Using JMP**.

Summary and Getting Help in JMP

In this chapter, we have provided an introduction to JMP, and have presented some methods for exploring and getting to know your data. We discussed common problems with data quality, and introduced a few methods for preparing data for modeling.

As you become familiar with JMP and the methods introduced in this book, you'll likely have many questions. Several forms of help are directly available from within JMP:

- The **help tool** or **question mark** on the toolbar: Select the question mark, and then click on an area of a data table or report on which you need assistance. Context-sensitive help tells you about the items located near the location of your click, and the index can be used to fine-tune your search.

- **Hover** help: Hover over a statistic in any window (in a clockwise direction in Windows), and JMP will provide a box with a definition of the statistic and help with interpretation.

- Help within dialog windows: The dialog windows of most analysis platforms include a **Help** button (or a question mark) that provides direct access to the JMP help files for the particular analysis.

- **JMP Help**: Accessed from the **Help** menu, this is the JMP searchable help system.

- **Books**: Also accessed from the **Help** menu, the JMP documentation is installed with JMP as searchable PDFs. The books are organized into the major topical areas. The last item in the list, **JMP Documentation Library**, opens one PDF file with all of the books. This allows you to search all of the documentation books at one time.

Since this book focuses on predictive modeling, you may also have questions regarding statistical concepts or topics not formally addressed in this book. An introductory statistics text (as well as numerous Internet resources) can be really handy to help fill in the gaps.

Exercises

Exercise 3.1: Use the **Equity.jmp** data set, from the sample data directory, for this exercise. Save this file using a new name (for example **EquityYourName.jmp**). We will return to this data in exercises in future chapters, so save scripts for your work to the data table, and save the file often.

a. Use the **Graph Builder** to re-create the graph in Figure 3.9, using **BAD** versus the other continuous predictors. Do any of the predictors appear to be related to **BAD**?

b. Use the **Graph Builder** to re-create the graph in Figure 3.10, using **BAD** versus **REASON**. Interpret this graph. Does there appear to be a relationship between **BAD** and **REASON**?

c. Use **Analyze > Fit Y by X** to analyze the relationship between **BAD (Y, Response)** and **LOAN** and **REASON (X, Factor)**. Explore the options under the red triangle for each of these analyses.

   i. Describe the relationship between **BAD** and **LOAN**.

   ii. Describe the relationship between **BAD** and **REASON**.

d. Use the **Distribution** platform to create a histogram and summary statistics for **DELINQ**, **VALUE**, and **MORTDUE**. Describe the shapes of these distributions.

e. Re-create the formula in Figure 3.21 to bin **DELINQ** into three groups.

   i. After creating the formula, use the **Distribution** platform to graph **DELINQ** and this new column. Then check your work (to make sure the binning was done correctly).

   ii. Later, we'll create a model to predict **BAD** from the available predictors. In this context, does binning **DELINQ** make sense? What impact will using the binned data (over the original variable) have on our model?

f. Refer to the distributions of **VALUE** and **MORTDUE** created in part d above. Use the **Graph Builder** and dynamic transformations (shown in Figure 3.20) to explore different transformations of these variables.

   i. Which transformations, if any, appear to normalize the variables? (Hint: use the **grabber** (the hand) tool to change the binning of the histogram in the **Graph Builder**. Click on the histogram and drag up or down to change the binning.)

   ii. From the context of modeling, explain why it might make sense to transform these variables.

Exercise 3.2: Use the **Bands Data.jmp** data set, from the sample data directory, for this exercise. This data set consists of 539 runs of a rotogravure printing press. The response, **Banding?**, indicates that there was a delay in the printing process due to a defect known as banding. Data were captured on 37 attributes. A team is tasked with identifying potential causes of banding, and will use exploratory and predictive modeling tools later in the project. Use the methods introduced in this chapter to explore and understand the data, and to prepare the data for modeling.

a. Other than **Banding?**, how many categorical variables are there? How many continuous variables are there?

b. Use the **Distribution** platform to create bar charts and frequency distributions for each of the categorical variables.

   i.   Which variables have several levels (categories)?

   ii. Which variables have most of the observations in only one category?

c. Use the **Distribution** platform to create histograms and summary statistics for the continuous variables.

   i.   Describe the shapes of the distributions for **ESA Voltage**, **ESA Amperage**, **Wax**.

   ii. Describe the distribution for **Chrome Content**. Explore the distribution (using your mouse and dynamic linking to the data table). What do you notice?

d. Identify two variables (one continuous variable and one categorical) that appear to be related to banding, and describe the nature of the relationships.

e. Identify 3 variables that do not appear to be related to banding, or do not appear to be informative. Explain.

f. Do any of the variables (not including **Banding?**) appear to be related to one another?

g. Is there a problem with missing values? Explain. Which variables are problematic, and how should this be addressed?

h. Are any of the variables messy or dirty? Explain. Which variables, in what way are they messy, and how should this be addressed?

References

Kutner, M., C. Nachtsheim, and J. Neter. 2004. *Applied Linear Statistical Models*, 4th ed. McGraw Hill.

SAS Institute Inc. 2015. *JMP 12 Multivariate Methods*. Cary, NC: SAS Institute Inc.

Sharpe, N., R. De Veaux, and P. Velleman. 2014. *Business Statistics*, 3rd ed. Pearson.

Shmueli, Galit, Nitin R. Patel, and Peter C. Bruce. 2010. *Data Mining for Business Intelligence: Concepts, Techniques, and Applications in Microsoft Ofice Excel with XLMiner*, 2nd ed. John Wiley & Sons, Inc.

Part 3

Model Building

In **Part III** we introduce four foundational modeling methods: **Multiple Linear Regression**, **Logistic Regression**, **Decision Trees** and **Neural Networks.** In each chapter we identify business use cases for the particular modeling method, take a look "under the hood" at some of the technical details behind the method, and provide two case studies involving application of the method to a business problem. Each chapter also includes a number of exercises.

Multiple Linear Regression

In the News

These days our entire lives revolve around predictions. Government departments project the cost of health exchanges, the rate of economic growth, next year's crop yields, the future birth rate and the arms buildup of unfriendly countries. Websites and retailers anticipate what we want to find and buy; oil companies gauge the best sites for drilling; pharmaceutical companies assess the probable efficacy of molecules on a disease; while, in the background, the bobble-heads on television incessantly spew out largely irrelevant and inaccurate forecasts. In the meantime, we busy ourselves with personal projections. How long will our commute take? When will the turkey be golden? How much will the price of a stock rise? What will the future value be of a law degree? (Michael Moritz."Are We All Being Fooled by Big Data?" January 3, 2013. Accessed at http://linkd.in/1zFuuq2.)

Representative Business Problems

Multiple linear regression is perhaps the most widely used and well-known statistical modeling tool. A straight-forward extension of simple linear regression, multiple regression is used to predict the average response value based on values of multiple predictors, or factors. For example, multiple regression can be used for:

- Identifying and optimizing critical to quality characteristics, with the goal of developing low cost and high quality products
- Predicting customer spending based on demographic information and historical buying patterns
- Developing pricing strategies based on product mix and consumer characteristics
- Establishing housing prices
- Determining the optimal timing for traffic lights to minimize traffic delays
- Predicting future product success from the results of a pilot study or trial

Preview of End Result

Suppose a model is developed to predict the amount of money a patron will spend on food per day while attending a popular professional golf tournament. A study is conducted and a model is developed based on household income and the average cost of food items sold:

Golf Tournament Daily Spend ($/day) = $25 + 0.08 * Annual Household Income (in thousands) – 0.21 * Average Cost of Food Items

So, if the average cost of food items is $10 and a patron's household income is $100,000, the predicted spending on food for that patron is:

$25 + (0.08 * 100) + (-0.21 * 10) = $30.90 per day.

Looking Inside the Black Box: How the Algorithm Works

Consider points on a scatterplot, where x is some predictor variable and y is some response. For example, the response in the previous section is money spent per day, and one of the x or predictor variables is household income. Now think about drawing a line that best represents or fits these data points. This line is our linear model fit to the data. The line provides a predicted value of y for each value of x.

The equation for this line can be expressed as:

$$y = b_0 + b_1 x$$

where b_0 is the y-intercept and b_1 is the slope. The y-intercept is the predicted value of y if the value of x is zero, and slope represents the change in y for every unit increase in x. Since we are fitting a model using data, this equation is generally expressed as:

$$\hat{y} = b_0 + b_1 x$$

where $\hat{y}$ represents the fitted value of y. This line doesn't fit the data perfectly. There is generally some difference between each response value and the line that we have drawn. This difference is called a residual, and it tells us how far the predicted value is from the observed value. This is illustrated in Figure 4.1. Each point is an observed value (the actual Daily Spend) for a given income level, and the vertical line tells us how far off each point is from the Daily Spend predicted by our linear model.

In fitting a line to the data, we are attempting to model the true unknown relationship between our predictor and our response. That is, we are trying to model reality. Our line is actually an estimate for the model. It represents the true unknown relationship between y and x, which is written as:

$$y = \beta_0 + \beta_1 x + \epsilon$$

Here, y is the response at a given value of x, β_0 is the true y-intercept, β_1 is the true slope, and ϵ represents random error.

Figure 4.1: Fitting a Line

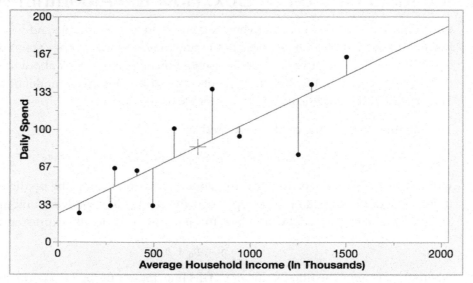

We estimate this true unknown model by drawing a line through our sample data that best fits the data. How do we measure "best"? Intuitively, the line that produces the smallest residual values is the best fit to our data. Because some residuals are positive and some are negative, we use an algorithm that finds the line with the smallest *sum of squared residuals*. This algorithm is aptly named the *Method of Least Squares*, and the line with the lowest sum of squared residuals, also called sum of squared errors, is referred to as the least squares regression line.

A visual representation of squared residuals is shown in Figure 4.2. Larger residuals have larger squared residuals, represented by pink squares. Smaller residuals have smaller squares. Intuitively, the least squares regression line is the line that results in the smallest squares in terms of total area.

Figure 4.2: Squared Residuals

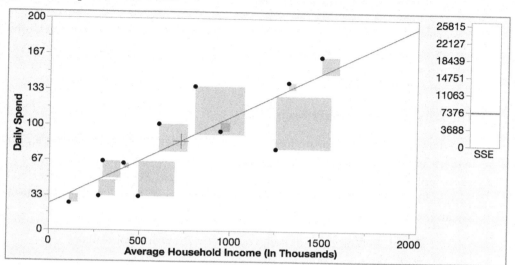

Simple linear regression, described above, involves one response and one predictor variable. In many situations, there are several variables that can be used to predict the response. In this case, we use the term multiple linear regression, and the formula is a straightforward extension of the simple regression model:

$$y = \beta_0 + \beta_i x_i + \cdots + \beta_k x_k + \epsilon$$

Each predictor has a corresponding slope, or coefficient, and the response is dependent upon the values of each predictor variable. In the original example, our model had two predictors, Annual Household Income (in thousands) and Average Cost of Food Items for Sale.

> *Note: Figures 4.1 and 4.2 were created with the* **Demonstrate Regression** *script for illustration purposes. The script is found in JMP (starting with JMP 12) under* **Help > Sample Data > Teaching Scripts > Interactive Teaching Modules**. *Click the* **Help** *button in the script for information on how to use the script to explore fitted lines, residuals, and sums of squares.*

Example 1: Housing Prices

A real estate company that manages properties around a ski resort in the United States wishes to improve its method for pricing homes. Sample data is obtained on a number of measures, including size of the home and property, location, age of the house, and a strength-of-market indicator.

The Data   HousingPrices.jmp

The data set contains information on about 45 residential properties near a popular North American ski resort sold during a recent 12-month period. The data set is a representative sample of the full set of properties sold during that time period (example provided by Marlene Smith, University of Colorado at Denver). The variables in the data set are:

> **Price**: Selling price of the property (in thousands of dollars)
>
> **Beds:** Number of bedrooms in the house
>
> **Baths:** Number of bathrooms in the house
>
> **Square Feet:** Size of the house in square feet
>
> **Miles to Resort:** Miles from the property to the downtown resort area
>
> **Miles to Base:** Miles from the property to the base of the ski resort's facing mountain
>
> **Acres:** Lot size in number of acres
>
> **Cars:** Number of cars that will fit into the garage
>
> **Years Old:** Age of the house at the time it was listed in years
>
> **DoM:** Number of days the house was on the market before it was sold

Applying the Business Analytics Process

Define the Problem

The real estate company wants to develop a model to predict the selling price of a home based on the data collected. The resulting pricing model will be used to determine initial asking prices for homes in the company's portfolio.

Prepare for Analysis

We begin by getting to know our data. As we saw in Chapter 3, we explore the distributions for each of our variables using **Analyze > Distribution**. We investigate relationships between the response and potential predictor variables using **Analyze > Multivariate Methods > Multivariate**.

Note that in this example, and throughout the modeling chapters, our focus is on particular modeling techniques. In each example, we use only a handful of methods, discussed in Chapter 3 to provide you with some familiarity with the data. However, we recommend that you use the graphical and numeric tools for exploring variables that were introduced, follow the suggestions for data preparation, and have a good understanding of each data set prior to modeling.

Exploring One Variable at a Time

Distribution output for all of the variables is shown in Figure 4.3a and Figure 4.3b. For each of these continuous variables, we see a histogram, a box plot, and various summary statistics.

Notice that the homes in this sample range in price from $160,000 to $690,000. Many of the homes have three or four bedrooms, two or three baths, are under 2,000 square feet on average, and are within twenty miles of the resort.

Figure 4.3a: A First Look at the Data – First 5 Housing Price Variables

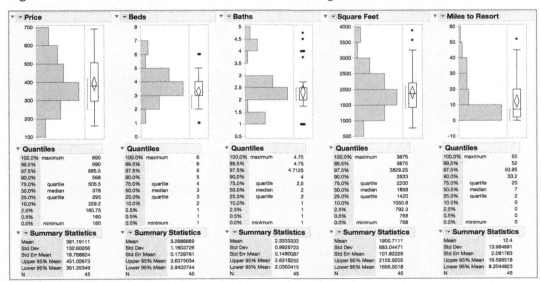

Figure 4.3b: A First Look at the Data – Last 5 Housing Price Variables

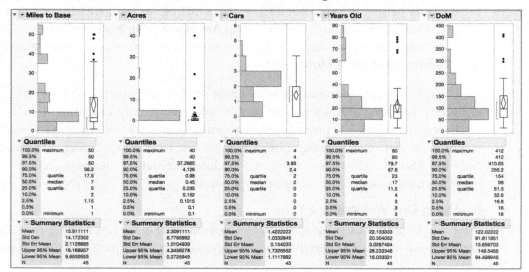

When we click on the $500-600,000 bin in the **Price** histogram, the values for these homes are also selected (shaded) in the other graphs. As one might expect, these more expensive homes tend to be on the larger side and are closer to the resort. For example, houses in this price range tend to have four or five bedrooms and three or four baths, are generally between 2000 and 3000 square feet, and are, for the most part, within twenty miles of the resort area.

Figure 4.4: Exploring Relationships

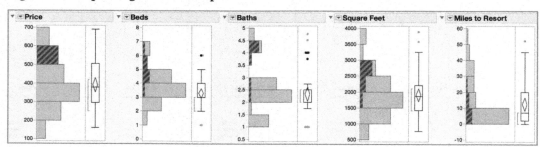

The histograms also tell us about the shapes of the distributions, and whether there are any patterns, unusual observations or potential outliers that may cause concern. **Miles to Resort** (in Figure 4.4) appears to be right-skewed, while the other four variables appear to be more symmetric. While the data set is small, there don't appear to be any unusually large or small values for any of the variables in Figure 4.4.

Exploring Many Variables at a Time

The **Multivariate** platform in JMP can be used for a more formal exploration of the relationships between the predictor variable and potential response variables. The **Correlations** table (Figure 4.5) shows correlations between all of the variables. Strong positive correlations are shown in blue in JMP, strong negative correlations are shown in red, and weak correlations are gray.

We're most interested in the correlation between **Price** and the other variables. We see strong positive correlations between **Price** and the number of beds, baths, and square feet, which are measures of house size, and negative correlations with miles from the base and miles from the resort, which are both distance measures.

We're also interested in understanding potential relationships between the predictor variables. For example, we can see strong correlations between each of the size measures (**Beds**, **Baths**, and **Square Feet**) and between the two distance measures (**Miles to Resort** and **Miles to Base**).

Figure 4.5: Correlations between Variables

	Price	Beds	Baths	Square Feet	Miles to Resort	Miles to Base	Acres	Cars	Years Old	DoM
Price	1.0000	0.6753	0.8001	0.6970	-0.5391	-0.6332	0.0251	0.4523	-0.3551	0.2298
Beds	0.6753	1.0000	0.7332	0.7282	-0.3509	-0.4241	-0.1473	0.1045	-0.3403	-0.0971
Baths	0.8001	0.7332	1.0000	0.7901	-0.3745	-0.4880	-0.1930	0.4357	-0.3267	0.1205
Square Feet	0.6970	0.7282	0.7901	1.0000	-0.1895	-0.2972	-0.1456	0.3728	-0.3037	-0.0110
Miles to Resort	-0.5391	-0.3509	-0.3745	-0.1895	1.0000	0.9480	0.2958	-0.1584	0.1082	-0.2219
Miles to Base	-0.6332	-0.4241	-0.4880	-0.2972	0.9480	1.0000	0.2634	-0.2612	0.2211	-0.1841
Acres	0.0251	-0.1473	-0.1930	-0.1456	0.2958	0.2634	1.0000	0.1474	-0.0295	0.3288
Cars	0.4523	0.1045	0.4357	0.3728	-0.1584	-0.2612	0.1474	1.0000	-0.2714	0.3137
Years Old	-0.3551	-0.3403	-0.3267	-0.3037	0.1082	0.2211	-0.0295	-0.2714	1.0000	0.0077
DoM	0.2298	-0.0971	0.1205	-0.0110	-0.2219	-0.1841	0.3288	0.3137	0.0077	1.0000

These relationships can be examined visually with the scatterplot matrix, which displays all of the two-way scatterplots between each pair of variables. Figure 4.6 shows the correlations and scatterplot matrix for three of the variables: **Price**, **Miles to Resort**, and **Miles to Base**. In the first row of the matrix, the *y*-axis for each of the graphs is **Price**, and the *x*-axis corresponds to the variable on the diagonal. So, for example, the two scatterplots in the first row display the relationship between **Price** and **Miles to Resort** and between **Price** and **Miles to Base**.

In each scatterplot, the correlation is displayed graphically as a density ellipse. The tighter (less circular) the ellipse, the stronger the correlation. The direction of the ellipse indicates whether the correlation is positive (the ellipse slopes up) or negative (the ellipse slopes down).

Looking at the variables, individually and together, helps us understand our data and potential relationships. We are starting to get a sense of the data and the variables that might need to be included in the model.

Note that other graphical tools, such as **Fit Y by X** (under the **Analyze** menu) and **Graph Builder** (under the **Graph** menu) can also be useful in exploring potential bivariate and multivariate relationships. We urge you to explore this data set on your own using all of these tools.

Figure 4.6: Visually Examining Correlations with a Scatterplot Matrix

Correlations

	Price	Miles to Resort	Miles to Base
Price	1.0000	-0.5391	-0.6332
Miles to Resort	-0.5391	1.0000	0.9480
Miles to Base	-0.6332	0.9480	1.0000

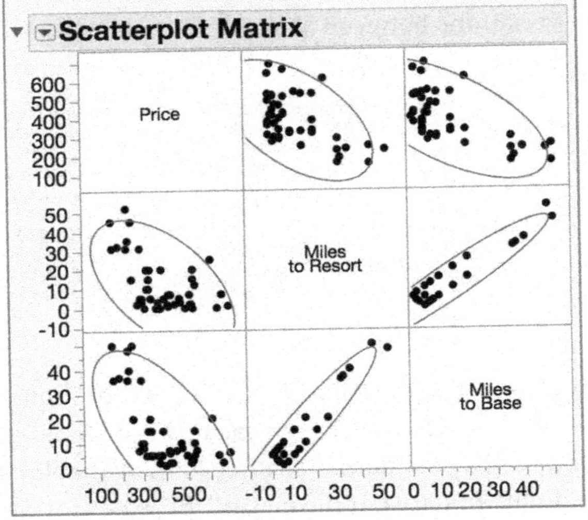

Build the Model

Our goal is to develop a model to predict the selling price of a home based on available data. Multiple linear regression is one of the core methods that can be used to develop a model to predict a continuous response from multiple predictor variables.

We begin by fitting a model using **Price** as the **Y** (response variable) and all of the potential factors as model effects using **Analyze > Fit Model**, as shown in Figure 4.7. Click **Run** to run the model.

Figure 4.7: Fit Model Dialog Window

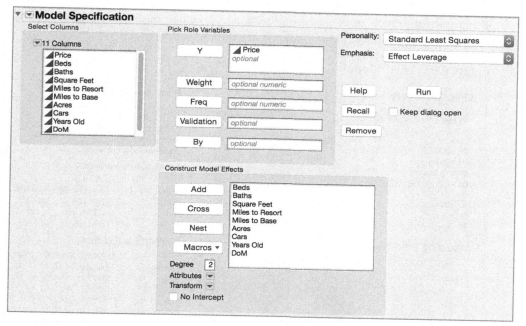

The results are shown in Figure 4.8. The **Effect Summary** table shows each of the terms in the model, sorted in ascending order of the *p*-value.

The **Actual by Predicted Plot** provides a graphical indication of the overall significance of the model. The closer the data points are to the diagonal line, the better our model does at explaining the variation in the response. (The solid diagonal line on this plot is the line where the *actual* value equals the *predicted* value.)

The **Summary of Fit** table provides key statistics, such as **RSquare** and **RSquare Adj** (adjusted R Square). RSquare indicates the percent of variation in the data that is explained by our model (0.81, or 81%). Because RSquare can be inflated simply by adding additional predictors to the model, the adjusted RSquare is sometimes used instead of RSquare for comparing models with more than one predictor (the "adjustment" applied to the **RSquare Adj** is based on the number of terms in the model).

The **Analysis of Variance** table indicates that the overall model is statistically significant. The *p*-value, reported as **Prob > F**, is < .0001.

The **Parameter Estimates** table provides coefficients for our model, along with *p*-values for each of the terms in the model.

Other output (effects tests, a residual plot, and leverage plots) is also provided by default, and additional options are available under the top red triangle.

Figure 4.8: Fitted Model

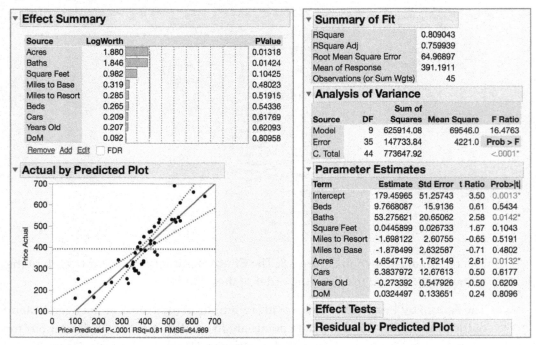

In Figure 4.8, we see that only two of the predictors (**Baths** and **Acres**) are significant at the 0.05 level, given that all of the other variables are in the model. Does this mean that none of the other predictors are important in predicting housing prices? We'll want to reduce this model to include only those variables that, in combination, do a good job of predicting the response. This is known as the principle of *parsimony:* the simplest model that can predict the response well is often the best model. Given that we have nine predictor variables, there can be many possible models to predict **Price**, from very simple models to more complex. In general, our goal is to find a concise model that makes sense, fits the data, and predicts the response well (Hansen, 2001). But, since significance is dependent upon which other variables are in the model, it is difficult to determine which terms to keep in the model and which to remove.

For illustration, we click **Remove** at the bottom of the **Effect Summary** table to remove **Baths** from the model (top, in Figure 4.9). **Acres** is no longer significant at $\alpha = 0.05$, but **Square Feet** is now significant (bottom, in Figure 4.9).

Figure 4.9: Effect Summary Table with and without Baths

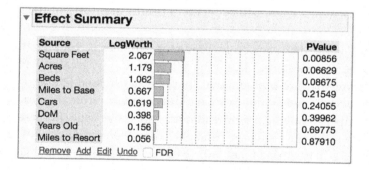

Part of the issue is that some of the variables are correlated with other variables in the model. Recall the correlation and scatterplot for **Miles to Resort** and **Miles to Base** (Figure 4.6). There is a very strong correlation between these two predictor variables. This means that they are somewhat redundant to one another. In fact, the resort and the base are in nearly the same geographic location.

A Bit About Multicollinearity

When two or more predictors are correlated with one another, the term *multicollinearity* is used. If multicollinearity is severe, then it is difficult to determine which of the correlated predictors are most important. In addition, the coefficients and standard errors for these coefficients may be inflated and the coefficients may have signs that don't make sense.

A measure of multicollinearity is the *VIF* statistic, or *Variance Inflation Factor*. The VIF for a predictor, VIF_j, is calculated using the following formula:

$$VIF_j = \frac{1}{1 - RSquare_{X_j}}$$

For each predictor, X_j, a regression model is fit using X_j as the response and all of the other X variables to predict X_j. The RSquare for that model fit $(RSquare_{X_j})$ is calculated, and is then used to calculate the VIF_j. An $RSquare_{X_j}$ of 0.9 results in a VIF_j of 10, while an $RSquare_{X_j}$ of 0.99 results in a VIF_j of 100.

If the VIF is 1.0, then each of the predictor variables is completely independent of the other predictor variables. But if the VIF is large (say, greater than 10), then the multicollinearity is a problem that should be addressed (Neter, 1996). In some cases, this can be resolved by removing a redundant term from the model. In more severe cases, simply removing a term will not address the issue. In these cases, variable reduction techniques such as Principal Components Analysis (PCA), Partial Least Squares (PLS), tree-based methods (covered in Chapter 6), and generalized regression methods (Chapter 9) are recommended.

In Figure 4.10 (on the left), we see VIFs for the original model. To display VIFs, right-click on the **Parameter Estimates** table and select **Columns > VIF**. The VIFs for most of the predictors are relatively small (< 5), while the VIFs for **Miles to Resort** and **Miles to Base** are both over 10, indicating that multicollinearity is a problem. Since we've learned that these two variables are largely redundant to one another, it makes sense to re-fit the model with only one of these variables. Subject matter knowledge can be used to determine which variable to remove. On the right in Figure 4.9, we see the results after removing **Miles to Resort** from the model. Notice that all of the VIFs are now low. In addition, **Miles to Base** is now significant, and the coefficient and the standard error for **Miles to Base** have both changed substantially!

Figure 4.10: Variance Inflation Factor, VIF

▼ **Parameter Estimates**

Term	Estimate	Std Error	t Ratio	Prob>|t|	VIF
Intercept	179.45965	51.25743	3.50	0.0013*	
Beds	9.7668087	15.9136	0.61	0.5434	3.5544519
Baths	53.275621	20.65062	2.58	0.0142*	4.3822145
Square Feet	0.0445899	0.026733	1.67	0.1043	3.4757465
Miles to Resort	-1.698122	2.60755	-0.65	0.5191	13.822325
Miles to Base	-1.878499	2.632587	-0.71	0.4802	14.510754
Acres	4.6547176	1.782149	2.61	0.0132*	1.5212784
Cars	6.3837972	12.67613	0.50	0.6177	1.7883544
Years Old	-0.273392	0.547926	-0.50	0.6209	1.2901804
DoM	0.0324497	0.133651	0.24	0.8096	1.5627487

▼ **Parameter Estimates**

Term	Estimate	Std Error	t Ratio	Prob>|t|	VIF
Intercept	181.44396	50.75587	3.57	0.0010*	.
Beds	11.370069	15.59576	0.73	0.4707	3.4693829
Baths	50.724242	20.11275	2.52	0.0162*	4.224488
Square Feet	0.0430186	0.026411	1.63	0.1121	3.4474345
Miles to Base	-3.493138	0.877935	-3.98	0.0003*	1.6400357
Acres	4.3625859	1.710918	2.55	0.0152*	1.4248944
Cars	5.4694553	12.49696	0.44	0.6642	1.766415
Years Old	-0.192613	0.529416	-0.36	0.7181	1.2240612
DoM	0.0592831	0.12612	0.47	0.6412	1.414216

To further assist in refining our model, after removing **Miles to Resort**, we'll rely on an automated variable selection approach, *stepwise regression*. We proceed with *stepwise regression* to identify the best subset of significant factors. **Stepwise** is a method, or a **Personality**, available in the **Fit Model** dialog (Figure 4.11).

Note that stepwise regression does not address multicollinearity. If correlated terms are used as inputs, stepwise may not result in the "best" model because variable selection will be determined by which variables are selected first.

Figure 4.11: Fit Model Stepwise Dialog

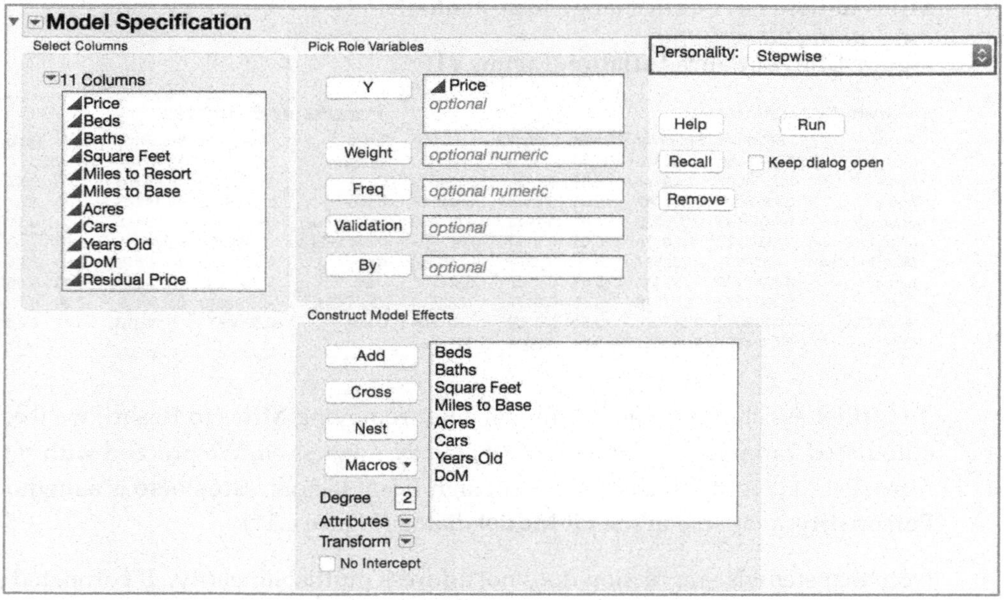

Stepwise regression provides a number of stopping rules for selecting the best subset of variables for the model. The default rule is **Minimum BIC**, or minimum *Bayesian Information Criterion*. The **Direction**, which is set to **Forward** by default, indicates that variables will be added to the model one at a time. After you click **Go**, the model with the smallest BIC statistic is selected.

Another common rule, which works in a similar manner, is **Minimum AICc** (*Akaike's Information Criterion*, with a correction for small sample sizes). Both of these rules attempt to explain the relationship between the predictors and the response, without building models that are overly complex in terms of the number of predictors. Since different criteria are used to determine when to stop adding terms to the model, these stopping rules may lead to different "best" models (Burnham, 2002).

We will develop a model using each criterion and then compare results. First, we use the default **Minimum BIC** criterion, and click **Go** to start the selection process. We identify four factors for the model: **Baths**, **Square Feet**, **Acres**, and **Miles to Base** (see Figure 4.12).

Figure 4.12: Stepwise Regression Variable Selection Using BIC

▼ ⊟ Stepwise Fit for Price

▼ Stepwise Regression Control

Stopping Rule: Minimum BIC ⬦ ➡ Enter All   Make Model

Direction: Forward ⬦   ⬅ Remove All   Run Model

[Go] [Stop] [Step]

	SSE	DFE	RMSE	RSquare	RSquare Adj	Cp	p	AICc	BIC
	153342.58	40	61.915785	0.8018	0.7820	1.9193847	5	507.9345	516.564

▼ Current Estimates

Lock	Entered	Parameter	Estimate	nDF	SS	"F Ratio"	"Prob>F"
✓	✓	Intercept	197.150171	1	0	0.000	1
☐	☐	Beds	0	1	1148.056	0.294	0.59063
☐	☑	Baths	59.2080973	1	46387.91	12.100	0.00123
☐	☑	Square Feet	0.05112326	1	19597.28	5.112	0.02927
☐	☑	Miles to Base	-3.7985104	1	90774.52	23.679	1.81e-5
☐	☑	Acres	5.00610917	1	46854.78	12.222	0.00117
☐	☐	Cars	0	1	311.3432	0.079	0.77968
☐	☐	Years Old	0	1	969.519	0.248	0.62118
☐	☐	DoM	0	1	408.8354	0.104	0.7485

Next, we change the stopping rule to **Minimum AICc**, click **Remove All** to clear the estimates from the BIC model, and then click **Go**. In this example, BIC and AICc yield the same set of factors (AICc output not shown). This isn't always the case.

We have identified a common set of factors for the model using two different stopping criteria. To run this regression model, select **Make Model**. Then in the **Fit Model** dialog, click **Run** (or, simply select **Run Model** from within the **Stepwise** platform).

Recall that our original model (Figure 4.8), with nine predictors, had only two significant terms and an adjusted R Square of 0.76. The results of fitting this reduced model are shown in Figure 4.13. As expected, this model is significant (Prob > F < .0001), and the four terms in the model are also significant. The adjusted R square is 0.782, which is slightly higher than our original model.

Figure 4.13: Revised Fitted Model

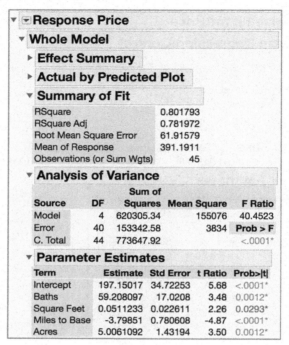

Before using this model, we need to check a few key assumptions. Namely, that our model errors are independent, have equal variance, and are normally distributed. Another key assumption is that the relationship between our response and the predictors is linear (i.e., that there isn't an underlying non-linear relationship).

The variation in the *residuals*, which is another word for the errors, shows us the variation in the response that could not be explained by the model that we have fit. Plots of residuals can be used to check that our assumptions about the model errors were correct. The default residual plot in JMP shows the residuals for each point plotted against predicted values. If our model assumptions are met, the points should be randomly scattered about the center line (zero), with no obvious pattern (just a cloud of seemingly random points). Other residual plots are also available (under the **red triangle > Row Diagnostics**), and the residuals can be saved (using **red triangle > Save Columns > Residuals**) and evaluated using **Distribution** or the **Graph Builder**.

The **Residual by Predicted** plot (see Figure 4.14) shows a somewhat curved pattern. That is, the largest residuals are at the lower and higher predicted values, while the smallest residuals are in the middle. This subtle pattern could be due to a term that is missing from the model. For example, the model may fit better if an interaction or quadratic (squared) term is added, or there may be an important variable that we've missed altogether. A *quadratic term* is used to explain curvature in the relationship between the factor and the response. An *interaction term* is used if the relationship between one factor and the response depends on the setting of another factor. We'll revisit interactions and quadratic terms in an exercise.

The pattern that we see in the residuals may also be due to outliers or influential observations, which can tilt or warp the regression model.

Figure 4.14: Examining Residuals

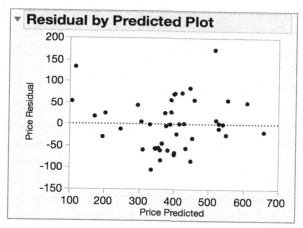

A statistic that helps us determine whether particular points are influencing the model is *Cook's D*, or *Cook's Distance*. Cook's D values for each observation can be saved to the data table, and then plotted using the **Analyze > Distribution** platform (see the plot of Cook's D values in Figure 4.15). To save Cook's D values from the Fit Model output window, click the red triangle and select **Save Columns > Cook's D Influence**.

A high Cook's D value for a particular observation indicates that the model predictions with and without that observation are different. What is considered high? A general rule of thumb is that any Cook's D value >1 is worthy of investigation (Cook, 1982). Observation #7, with a Cook's D value over 6, has a large influence on our model.

Figure 4.15: Cook's D Values

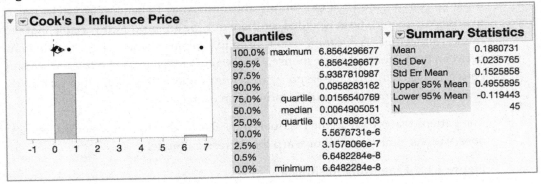

To illustrate how an influential point can impact model, see Figure 4.16. We use **Analyze > Fit Y by X**, with **Price** as **Y, Response** and **Acres** as **X, Factor**, and fit a line (select **Fit Line** from the red triangle). Then, we exclude observation 7, and again select **Fit Line**. The resulting regression lines are labeled (using the **Annotate** tool from the toolbar in JMP).

Note the difference in the slopes for fitted regression lines with and without observation 7 included in the model! Clearly, these two models will result in different predicted values, particularly for properties with higher acreage.

Figure 4.16: Illustration of Influential Point

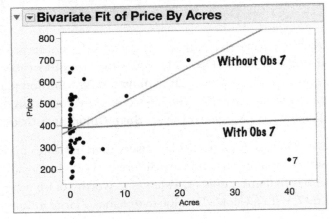

Upon investigation, we find that this property is actually a 40-acre farm rather than a residential property. The focus of this pricing study is residential properties. Since this property is not of direct interest in this study and is influencing our predictions, we exclude and hide observation 7 (use **Rows > Hide and Exclude**) and proceed without it.

Since the data set is small and stepwise was performed using observation 7, the procedure may result in a different reduced model if run without this observation. We return to stepwise, and find that the same reduced model is produced without this observation (Figure 4.17).

Compared to the model in Figure 4.13, this model has a higher R Square Adjusted (0.84), and most of the *p*-values for the terms in the model are lower. In particular, the *p*-value for **Acres** has dropped from 0.0012 to <.0001 (and the coefficient is now much larger). In addition, the residuals now appear more randomly scattered about the center line, with no obvious patterns, evidence of curvature, or lack of constant variance.

Figure 4.17: Model and Residuals after Removing Influential Point

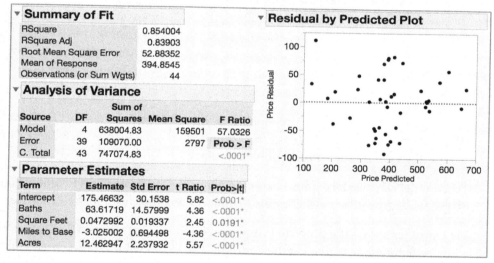

Given that the houses are scattered around the geographic area surrounding the resort and there is no obvious clustering of points in the residual plot, we have some additional comfort that the independence assumption is also met.

For additional confirmation that the normality assumption has been met, we can save the residuals to the data table (click the red triangle and select **Save Columns > Residuals**), and then use a histogram and normal quantile plot to check normality (use **Distribution**, and select **Normal Quantile Plot** from the red triangle). We leave it to the reader to confirm that the normality assumption has indeed been met.

Satisfied with our final model, we can now use this model to predict home prices. The terms in our model, and their coefficients, are given in the **Parameter Estimates** table (left, in Figure 4.18). Each estimate tells us how much the predicted selling price changes with a change in the value of the predictor.

The parameter estimates can be rewritten as a formula, or prediction expression (right, in Figure 4.18). To calculate the selling price of a home, we simply need to plug in the number of baths, the square feet, the miles to base, and the acres into the formula.

Figure 4.18: Parameter Estimates and Prediction Expression

▼ Parameter Estimates					▼ Prediction Expression
Term	**Estimate**	**Std Error**	**t Ratio**	**Prob>|t|**	175.466319174148
Intercept	175.46632	30.1538	5.82	<.0001*	
Baths	63.61719	14.57999	4.36	<.0001*	+ 63.6171900658549 * Baths
Square Feet	0.0472992	0.019337	2.45	0.0191*	+ 0.04729922796845 * Square Feet
Miles to Base	-3.025002	0.694498	-4.36	<.0001*	+ -3.025002137271 * Miles to Base
Acres	12.462947	2.237932	5.57	<.0001*	+ 12.4629468282909 * Acres

To display the prediction expression, click the red triangle and select **Estimates > Show Prediction Expression**. To save this formula to the data table, click the red triangle and select **Save Columns > Prediction Formula**.

The model can also be explored graphically using the **Prediction Profiler** (Figure 4.19). To access the profiler, select **Factor Profiling > Profiler** from the red triangle.

The profiler shows the predicted response (on the far left) at specified values of each of the predictor values (given at the bottom). The initial values for the predictors are predictor averages, and vertical red lines are drawn at these values. The starting value for the response is also the overall average (the mean **Price** in this example), and the bracketed values are the 95% confidence interval for the average. The confidence interval can be used to determine a *margin of error* for the prediction. The margin of error is the *half-width* of the confidence interval, or half the range of the interval. In this example, the margin of error is approximately (410.984 - 378.732)/2 = $16.125K.

Drag the vertical lines for a predictor to change the value for that predictor. The slopes of the lines for each predictor indicate whether predicted **Price** will increase or decrease if the predictor value increases, assuming that the other predictor values are held constant.

Figure 4.19: Using Prediction Profiler

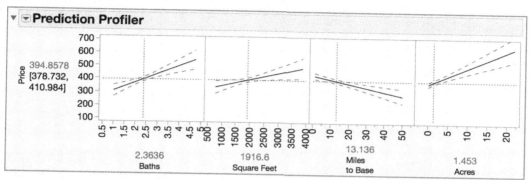

Summary

In this example, we have created a regression model for home selling prices using historical data and have assessed model assumptions to ensure that the model makes sense. If we are satisfied with our model's suitability and performance, we can put the model to use to predict selling prices of new homes entering the market in this geographic region. For example, our model tells us that the predicted selling price of a 2500 square foot home with 3 baths that is 13 miles to the base and sits on one acre is just over $457,704, with a margin of error of approximately $21.5K. (You can confirm this by entering these values into the **Prediction Profiler**.)

Of course, we might ask if this is the best possible model to predict selling price. Can we build a better model? Our margin of error is relatively large, and our standard deviation (reported as Root Mean Square Error in Figure 4.17) is just under $53K. Can we develop a model that provides more precise predictions? What if we built a model using information on a larger sample of houses? What if we include additional information for each of the houses sold, such as recent renovations, measures of home quality, or the time of year that the home was first put on the market? Would this lead to a better model?

We should also keep in mind that all models need to be updated periodically. Housing prices change over time, so the model to predict housing prices should be updated to stay current and reflect these changes.

Example 2: Bank Revenues

A bank wants to understand how customer banking habits contribute to revenues and profitability. The bank has the customer age and bank account information, such as whether the customer has a savings account, if the customer has received bank loans, and other indicators of account activity.

The Data   BankRevenue.jmp

The data set contains information on 7420 bank customers:

> **Rev_Total:** Total revenue generated by the customer over a 6-month period.
>
> **Bal_Total:** Total of all account balances, across all accounts held by the customer.
>
> **Offer:** An indicator of whether the customer has received a special promotional offer in the previous one-month period. Offer=1 if the offer was received, Offer=0 if it was not.
>
> **AGE:** The customer's age.
>
> **CHQ:** Indicator of debit card account activity. CHQ=0 is low (or zero) account activity, CHQ=1 is greater account activity.
>
> **CARD:** Indicator of credit card account activity. CARD=0 is low or zero account activity, CARD=1 is greater account activity.
>
> **SAV1:** Indicator of primary savings account activity. SAV1=0 is low or zero account activity, SAV1=1 is greater activity.
>
> **LOAN:** Indicator of personal loan account activity. LOAN=0 is low or zero account activity, LOAN=1 is greater activity.
>
> **MORT:** Indicator of mortgage account tier. MORT=0 is lower tier and less important to the bank's portfolio. MORT=1 is higher tier and indicates the account is more important to the bank's portfolio.
>
> **INSUR:** Indicator of insurance account activity. INSUR=0 is low or zero account activity, INSUR=1 is greater activity.
>
> **PENS:** Indicator or retirement savings (pension) account tier. PENS=0 is lower balance and less important to bank's portfolio. PENS=1 is higher tier and of more importance to the bank's portfolio.

Check: Indicator of checking account activity. Check=0 is low or zero account activity, Check=1 is greater activity.

CD: Indicator of certificate of deposit account tier. CD=0 is lower tier and of less importance to the bank's portfolio. CD=1 is higher tier and of more importance to the bank's portfolio.

MM: Indicator of money market account activity. MM=0 is low or zero account activity, MM=1 is greater activity.

Savings: Indicator of savings accounts (other than primary) activity. Savings=0 is low or zero account activity, Savings=1 is greater activity.

AccountAge: Number of years as a customer of the bank.

Applying the Business Analytics Process

Define the Problem

We want to build a model that allows the bank to predict profitability for a given customer. A surrogate for a customer's profitability that is available in our data set is the **Total Revenue** a customer generates through their accounts and transactions. The resulting model will be used to forecast bank revenues and guide the bank in future marketing campaigns.

Prepare for Modeling

We begin by looking at the variable of interest, total revenue (**Rev_Total**) using **Graph > Graph Builder**. **Rev_Total** is highly skewed, which is fairly typical of financial data (Figure 4.20).

> *Note: To explore the underlying shape of the distribution, select the **Grabber** (hand) tool from your toolbar, click on the graph and drag up and down.*

Figure 4.20: Distribution of Total Revenue

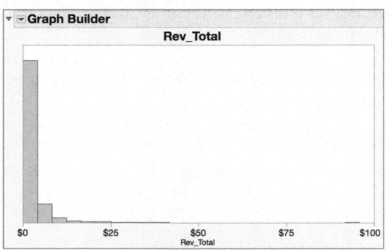

In regression situations, highly skewed data can result in a poorly fitting model. A transformation that can often be used to normalize highly skewed data, where all of the values are positive, is a log (natural logarithm) transformation (see Ramsey and Shafer, 2002, page 68).

We apply a log transformation to the **Rev_Total** variable directly in the **Graph Builder** and reexamine the distribution (Figure 4.21). To apply this transformation, right-click on the variable in the variable selection list, and select **Transform > Log**. Then, to save the transformation to the data table, right-click on **Log(Rev_Total)** and select **Add to Data Table**.

This transformation gives us a much less skewed and more symmetric distribution, so we use **Log(Rev_Total)** for the rest of our analysis.

A similar examination of the total account balance (**Bal_Total**), which also has a skewed distribution, leads to using the **Log(Bal_Total)** in our analyses.

Figure 4.21: Transformed Total Revenue Using Log Transformation

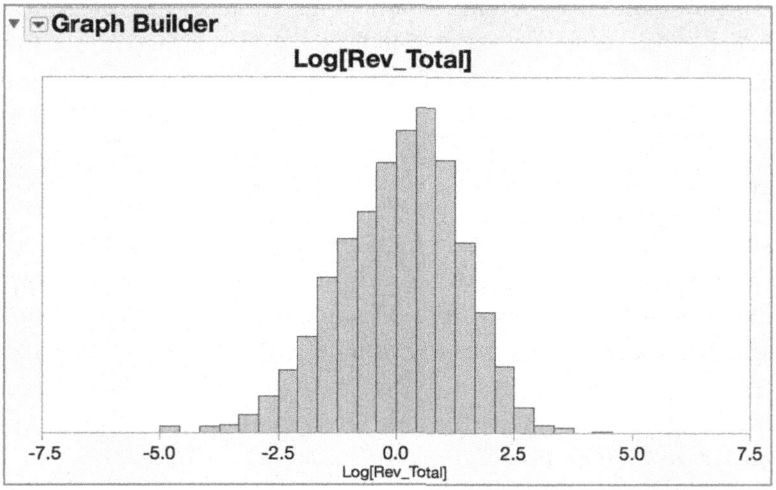

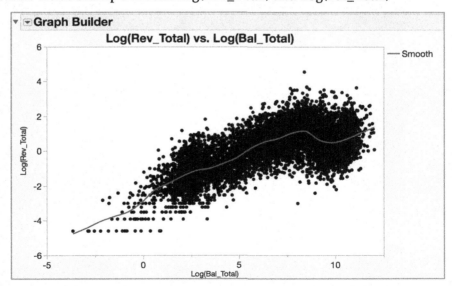

The relationship between the log total revenue and log total account balance is shown in the scatterplot in Figure 4.22. The relationship appears nearly linear at lower account balances—higher account balances generally have higher revenues. But the relationship seems to change at the higher account balances.

Figure 4.22: Relationship between Log(Rev_Total) and Log(Bal_Total)

Now we examine the other variables. We can see their distributions and also their relationship to **Log(Rev_Total)**. Many of the variables are categorical, with two-levels. Higher revenue values are selected in Figure 4.23, and we can see this selection across the other variables in our data set. (The **Arrange In Rows** option under the red triangle was used to generate Figure 4.23; not all variables are displayed.) Other than total account balance, **Log(Bal_Total)**, there is no variable that stands out as being strongly related to revenue.

As we have discussed, other graphical and analytic tools can be used to understand the data and explore potential relationships, such as **Fit Y by X** and **Graph Builder**. In addition, the **Data Filter** (under the **Rows** menu) and **Column Switcher** (under the **red triangle > Scripts** in any output window) are dynamic tools that allow you to dive deeper into your data to explore variables of interest and potential relationships. Again, we encourage you to explore the data using these tools on your own. See Chapter 3 for discussion and illustration of different exploratory tools.

> *Note: Recall that within JMP there are a number of preferences that can be set (under **File > Preferences** or **JMP > Preferences** on a Mac), and all JMP output is customizable with your mouse and keystrokes. Going forward, we periodically resize graphs and change axis scaling to better fit content on the page, and change marker sizes or colors to improve interpretability. We also turn off shaded table headings in output to provide a cleaner display (within **Preferences, Styles > Report Tables**).*

Figure 4.23: Relationships between Transformed Variables and Other Variables

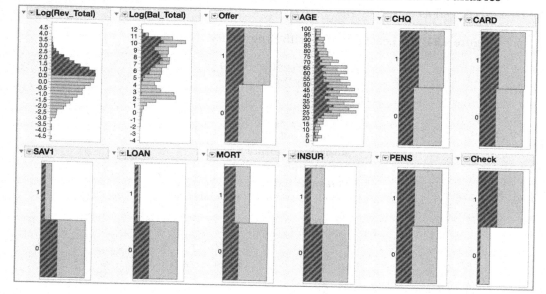

Build the Model

We now build a regression model to predict **Log(Rev_Total)** using **Fit Model** and the 15 potential predictor variables. There are some immediate signs of trouble (Figure 4.24). At the top of the **Fit Least Squares** window, we see some unexpected output, *Singularity Details*. This means that there are linear dependencies between the predictor variables. The first row of this table, **LOAN[0] = CD[0]** , indicates that JMP can't tell the difference between these two variables, **LOAN** and **CD**. The second line indicates that JMP can't tell the difference between **INSUR**, **MM**, and **Savings**.

The cause of this problem is illustrated in the **Distribution** output in Figure 4.25. The distributions of these three variables are identical. Every time **LOAN** = 1 (a customer has high loan activity), **MM** and **Savings** are also 1 (money market and savings activity are also high). The variables within each grouping are completely redundant to one another!

The result of this problem is seen in the parameter estimates table. JMP can't estimate all of these coefficients, indicating that the estimates for **LOAN** and **INSUR** are *biased*, and the estimates for **CD**, **MM**, and **Savings** are *zeroed*. JMP can estimate some of the parameters for the redundant variables (these estimates are biased), but not all (these are zeroed). Whether the variables appear as biased or zeroed depends entirely on the order in which they were entered into the model—those entered first into the model are displayed as biased.

Figure 4.24: Fit Least Squares with Singularity

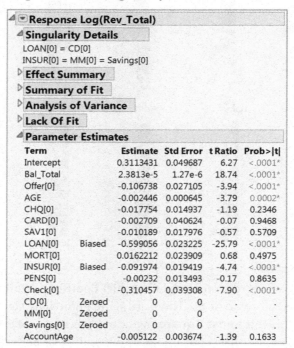

Figure 4.25: Distributions of INSUR, MM, and Savings

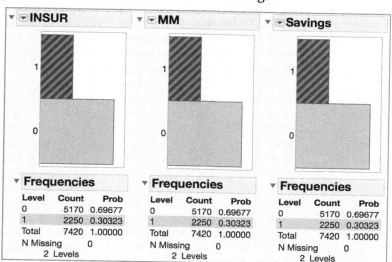

We refit the model without the redundant variables. In JMP 12, this can be done using the **Remove** button at the bottom of the **Effect Summary** table. We keep **LOAN** (and eliminate **CD**), and keep **INSUR** (eliminating **MM** and **Savings**). Note that this was an arbitrary decision: subject matter knowledge should guide the decision as to which redundant variables to remove (and which variables to keep in the model). As we remove each variable (or term), the **Singularity Details** table updates, along with all of the other statistical output. JMP is now able to estimate coefficients for each of the parameters (Figure 4.26).

Figure 4.26: Fit Least Squares Parameter Estimates without Singularity, Showing VIFs

◢ **Parameter Estimates**

Term	Estimate	Std Error	t Ratio	Prob>|t|	VIF
Intercept	-2.531352	0.044361	-57.06	<.0001*	.
Log(Bal_Total)	0.4421894	0.004931	89.68	<.0001*	2.539337
Offer[0]	-0.069263	0.019212	-3.61	0.0003*	4.1313113
AGE	-0.00057	0.000455	-1.25	0.2103	1.0364354
CHQ[0]	-0.004403	0.010521	-0.42	0.6756	1.2500319
CARD[0]	-0.783241	0.027512	-28.47	<.0001*	8.5478082
SAV1[0]	0.0109566	0.012739	0.86	0.3898	1.1167727
LOAN[0]	0.0587778	0.018132	3.24	0.0012*	1.4725238
MORT[0]	0.0160545	0.016929	0.95	0.3430	3.0203293
INSUR[0]	0.0371717	0.013774	2.70	0.0070*	1.8111374
PENS[0]	-0.000349	0.009562	-0.04	0.9709	1.0309563
Check[0]	0.6864921	0.028624	23.98	<.0001*	6.3844217

A quick check of the VIFs indicates that multicollinearity is not a serious issue (Figure 4.26).

Since we have 11 remaining potential predictor variables, we use again stepwise regression to help with variable selection. We use the **Stepwise Personality** in the **Fit Model** platform, with **Log(Rev_Total)** as our Y and the other variables as model effects. For this example, we use the **Minimum AICc** stopping rule.

Stepwise selects six variables for the model. These are checked under **Current Estimates** in Figure 4.27. Note that when using AICc (or BIC), the resulting models may include terms that are not significant. This is because both AICc and BIC build models based on *important effects* (effects that explain the relationship between the response and the predictors) rather than searching for *significant effects* (see Burhnam, 2002). However, in this example, all six selected variables have low *p*-values.

Figure 4.27: Stepwise Regression Dialog with Model Variables Selected

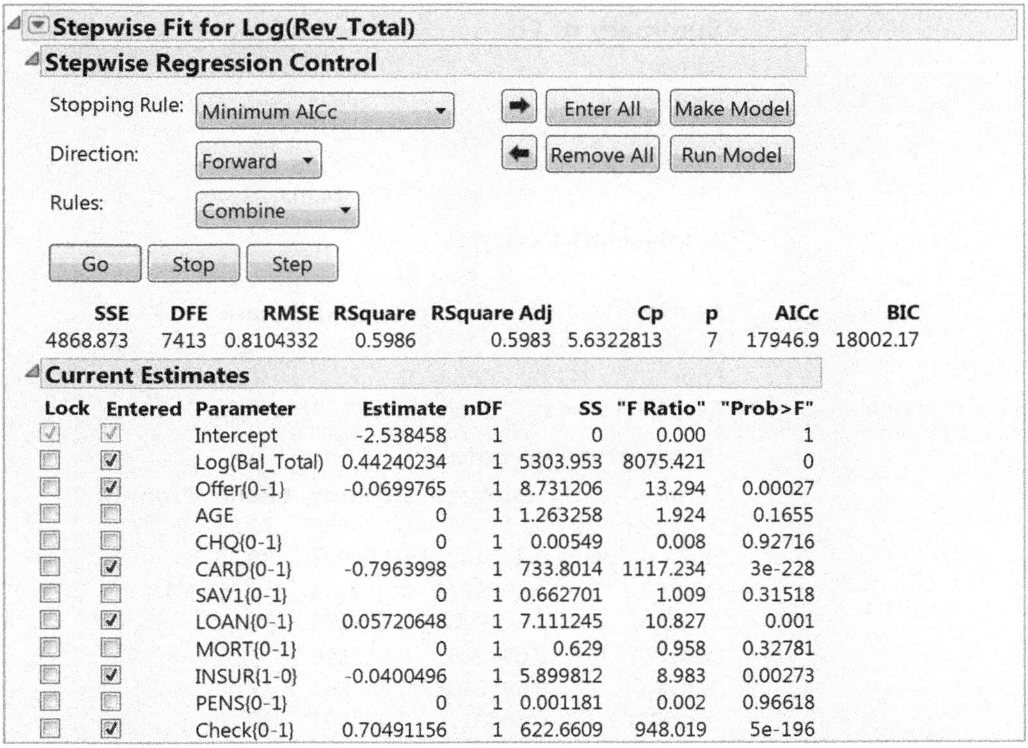

Stepwise Fit for Log(Rev_Total)

Stepwise Regression Control

Stopping Rule: Minimum AICc ▾   ➡ Enter All   Make Model

Direction: Forward ▾   ⬅ Remove All   Run Model

Rules: Combine ▾

| Go | Stop | Step |

SSE	DFE	RMSE	RSquare	RSquare Adj	Cp	p	AICc	BIC
4868.873	7413	0.8104332	0.5986	0.5983	5.6322813	7	17946.9	18002.17

Current Estimates

Lock	Entered	Parameter	Estimate	nDF	SS	"F Ratio"	"Prob>F"
✓	✓	Intercept	-2.538458	1	0	0.000	1
☐	✓	Log(Bal_Total)	0.44240234	1	5303.953	8075.421	0
☐	✓	Offer{0-1}	-0.0699765	1	8.731206	13.294	0.00027
☐	☐	AGE	0	1	1.263258	1.924	0.1655
☐	☐	CHQ{0-1}	0	1	0.00549	0.008	0.92716
☐	✓	CARD{0-1}	-0.7963998	1	733.8014	1117.234	3e-228
☐	☐	SAV1{0-1}	0	1	0.662701	1.009	0.31518
☐	✓	LOAN{0-1}	0.05720648	1	7.111245	10.827	0.001
☐	☐	MORT{0-1}	0	1	0.629	0.958	0.32781
☐	✓	INSUR{1-0}	-0.0400496	1	5.899812	8.983	0.00273
☐	☐	PENS{0-1}	0	1	0.001181	0.002	0.96618
☐	✓	Check{0-1}	0.70491156	1	622.6609	948.019	5e-196

We now run this model, and explore the results (Figure 4.28). As expected, the overall model is significant with a p-value < .0001, as are all of the terms in the model. The R Square is 0.5986, indicating that our model explains nearly 60% of the variation in the response.

Figure 4.28: Model Results, Reduced Model

Summary of Fit

RSquare	0.598624
RSquare Adj	0.598299
Root Mean Square Error	0.810433
Mean of Response	0.059558
Observations (or Sum Wgts)	7420

Analysis of Variance

Source	DF	Sum of Squares	Mean Square	F Ratio
Model	6	7261.582	1210.26	1842.661
Error	7413	4868.873	0.66	**Prob > F**
C. Total	7419	12130.455		<.0001*

Parameter Estimates

Term	Estimate	Std Error	t Ratio	Prob>|t|
Intercept	-2.538458	0.034835	-72.87	<.0001*
Log(Bal_Total)	0.4424023	0.004923	89.86	<.0001*
Offer[0]	-0.069977	0.019193	-3.65	0.0003*
CARD[0]	-0.7964	0.023826	-33.43	<.0001*
LOAN[0]	0.0572065	0.017386	3.29	0.0010*
INSUR[0]	0.0400496	0.013363	3.00	0.0027*
Check[0]	0.7049116	0.022894	30.79	<.0001*

Before interpreting the results of the regression model, we check that the regression assumptions are met. Since the data are from over 7400 different customers, we have some assurance that the independence assumption is met. The default residual versus predicted value plot (Figure 4.29) shows some diagonal striations in the lower left corner.

Figure 4.29: Residual versus Predicteds

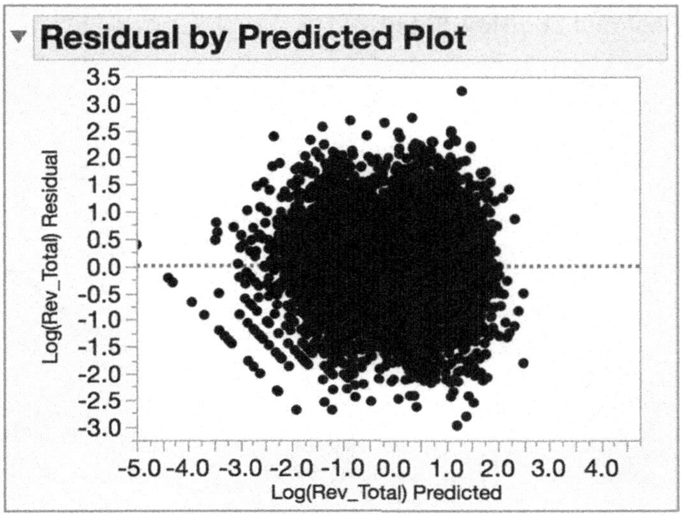

To explore these values, we use the **lasso tool** on the toolbar to select the observations, activate the data table, and then use the **F7** function key to scroll through these selected observations in the data table. The first strip on the left corresponds to revenue $0.01, and the second is revenue $0.02. This is the result of the fact that there are many customers who generate little, if any, revenue for the bank.

Otherwise, points appear randomly scattered around the center line (zero), and the residual plot shows no obvious evidence of unusual patterns.

For further exploration of the regression assumptions, we save the residuals to the data table (under the red triangle, select **Save Columns > Residuals**), and use the **Distribution** platform to generate a histogram with a normal quantile plot (Figure 4.30). These plots provide evidence that the normality assumption has been met.

Figure 4.30: Distribution of Residuals with Normal Quantile Plot

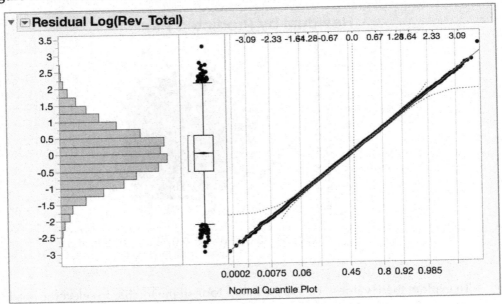

We also see (in Figure 4.30) that there are no serious outliers. A quick peek at Cook's D values (Figure 4.31) confirms that there are no highly influential observations. No single point is exerting too much influence over our model.

Figure 4.31: Checking Assumptions with Cook's D

Cook's D Influence Log(Rev_Total)

Quantiles			Summary Statistics	
100.0%	maximum	0.0040971292	Mean	0.0001397
99.5%		0.0018369614	Std Dev	0.0002799
97.5%		0.0008889624	Std Err Mean	3.2498e-6
90.0%		0.0003486082	Upper 95% Mean	0.0001461
75.0%	quartile	0.0001457648	Lower 95% Mean	0.0001334
50.0%	median	0.0000460879	N	7420
25.0%	quartile	0.0000103113		
10.0%		1.5692371e-6		
2.5%		8.4912948e-8		
0.5%		5.9889596e-9		
0.0%	minimum	5.223969e-12		

After investigating residuals and looking at Cook's D values, we have confidence that the regression assumptions have been satisfied. Our final model, shown in Figure 4.32, includes the following variables:

- The total account balance (**Log(Bal_Total)**)
- Whether the customer received a promotional offer (**Offer**)
- Credit card activity (**CARD**)
- Personal loan account activity (**LOAN**),
- Insurance account activity (**INSUR**)
- Checking account activity (**Check**)

All of the significant variables except **Log(Bal_Total)** are binary categorical variables. For the continuous predictor, **Log(Bal_Total)**, the coefficient in the parameter estimates table (top in Figure 4.32) indicates how the revenues change as the account balance changes. The positive coefficient indicates that revenues increase on average as account balances increase. The coefficient value itself is a little difficult to interpret because it reflects the transformation of **Rev_Total** to **Log(Rev_Total)**.

For each of the two-level categorical predictors, the parameter estimates show how the average response changes at the low level of each predictor. For example, the coefficient for **CARD[0]** is negative 0.7964. This indicates that log revenues are 0.7964 lower on average if credit card activity is low, and 0.7964 higher on average if credit card activity is high. The coefficients for **LOAN**, **Check**, and **INSUR** are all positive, indicating that low activity in these three accounts leads to higher revenues.

> **Note:** When fitting regression models in JMP, two-level categorical predictors are automatically transformed into coded indicator variables using a -1/+1 coding scheme. The parameter estimate is reported for the lowest level or value of the predictor. In this example, **CARD** is a nominal predictor with levels with 0 and 1. The term in the reduced model is represented as **CARD[0]**, and the parameter estimate is -0.7964 (see Figure 4.32). The estimate for **CARD[1]**, which is not reported, is +0.7964. To display both estimates, select **Expanded Estimates** from the **top red triangle > Estimates**.
>
> Many statistical software packages require dummy coding of categorical predictors, using a 0/1 "dummy" or "indicator" coding scheme. This is done prior to fitting the model, and results in different parameter estimates and a different interpretation of the estimates. For example, the parameter estimate for **CARD**, using 0/1 dummy coding, is 1.5928 instead of -0.7964. The sign is different, and the estimate is exactly

twice the magnitude. To confirm this, change the modeling type for **CARD** to **Continuous** (to tell JMP to use dummy coding) and refit the reduced model shown in Figure 4.28. Note that, although the parameter estimates are different, the two coding schemes produce identical model predictions.

To view the indicator-coded version of the parameter estimates in the **Fit Least Squares** output, select **Indicator Parameterization Estimates** from the **top red triangle > Estimates**. Further details of how JMP transforms categorical factors can be found in the Statistical Details section of the book Fitting Linear Models (under **Help > Books**).

The prediction profiler (bottom of Figure 4.32) can help us see the impact of changes in values of the predictor variables on **Log(Rev_Total)**.

Figure 4.32: Exploring the Reduced Model with the Prediction Profiler

Parameter Estimates

Term	Estimate	Std Error	t Ratio	Prob>|t|
Intercept	-2.538458	0.034835	-72.87	<.0001*
Log(Bal_Total)	0.4424023	0.004923	89.86	<.0001*
Offer[0]	-0.069977	0.019193	-3.65	0.0003*
CARD[0]	-0.7964	0.023826	-33.43	<.0001*
LOAN[0]	0.0572065	0.017386	3.29	0.0010*
INSUR[0]	0.0400496	0.013363	3.00	0.0027*
Check[0]	0.7049116	0.022894	30.79	<.0001*

Prediction Profiler

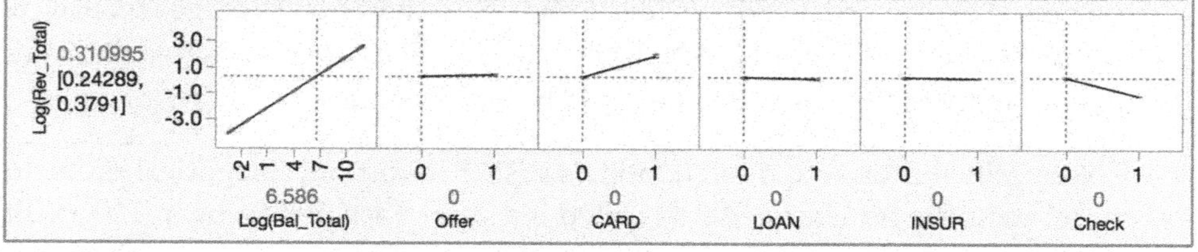

Clearly, **Log(Bal_Total)** has a large positive effect on the response. Three predictors, **Offer**, **LOAN**, and **INSUR**, while significant, have a relatively small effect on the response.

To show the predicted values for each bank customer, the prediction equation (the formula) can be saved to the data table (red triangle, **Save Columns > Prediction Formula**). Unfortunately, these are the log predicted values, which are difficult to interpret.

The inverse transformation (in this case the *exponential*, or Exp function) can be used to see the predicted values on the original scale. To apply this transformation, create a new column in the data table (we've named this column **Pred Rev_Total**). Then, right-click on the column and select **Formula** to open the **Formula Editor**, and use the **Transcendental > Exp** function from the **Functions (grouped)** list (see Figure 4.33).

Note that this formula can be created using a shortcut. Simply right-click on the saved prediction formula column, and select **New Formula Column > Transform > Exp**. JMP will create the new column with the stored formula shown in Figure 4.33.

Figure 4.33: Transforming Predicted Log(Rev_Total) to Predicted Rev_Total

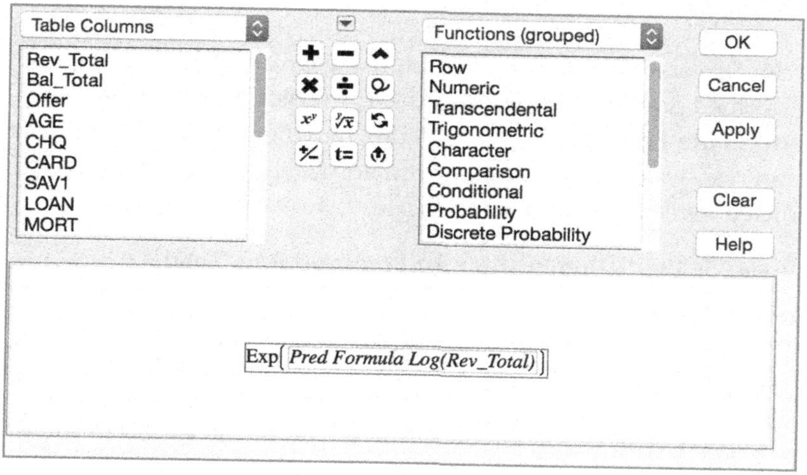

Now, we can explore the distribution of these values using **Distribution** or **Graph Builder** (Figure 4.34).

Figure 4.34: Distribution of Predicted Rev_Total

▼ Pred Rev_Total

Quantiles			Summary Statistics	
100.0%	maximum	12.218379622	Mean	1.580706
99.5%		6.5200938162	Std Dev	1.3093443
97.5%		4.9030345728	Std Err Mean	0.0152003
90.0%		3.2604062067	Upper 95% Mean	1.6105029
75.0%	quartile	2.1543797924	Lower 95% Mean	1.5509091
50.0%	median	1.4049795287	N	7420
25.0%	quartile	0.4411949596		
10.0%		0.2678533829		
2.5%		0.1509048307		
0.5%		0.0667292746		
0.0%	minimum	0.0067823167		

We can also explore the formula itself using **Graph > Profiler** (Figure 4.35). Select the transformed prediction formula as the **Y, Prediction Formula**, and check the **Expand Intermediate Formulas** box to drill down to the original saved prediction formula. Now, we can readily see and explore the impact of changes to each of the variables on the predicted revenues in the original scale.

Figure 4.35: Prediction Profiler for Predicted Rev_Total

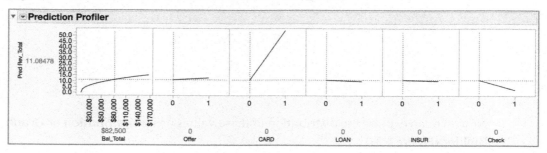

Summary

It is clear that high account balance customers, and those who use their credit card frequently, generate more revenue. What is curious is that high checking account usage seems to indicate lower revenue, and that customers with higher activity on the loan and insurance accounts have lower predicted revenue on average.

Was the promotional offer was successful? That is, did it lead to increased revenue? For a customer maintaining an account balance of $82,500, with low credit card, loan, insurance, and checking account activity, the promotional offer increased revenues from $11.08 to $12.75 on average. If this same customer had high credit card activity instead of low, the predicted revenue increased from $54.5 to $62.7. However, this analysis does not determine return on investment. Further information would need to be gathered to determine the cost of the promotional offer program and to examine the increased revenue relative to that cost. All of these insights lead to more questions, with new business problems to solve.

Exercises

Exercise 4.1: In this exercise, we use the **HousingPrices.jmp** data. In this chapter, we built a predictive model for **Price**, but limited model terms to main effects (that is, the predictors themselves). This is due, in part, to the fact that our data set is very small. However, other possible model effects include interactions and squared terms (quadratic effects).

Consider a model with all of the original predictors, plus one two-factor interaction and one quadratic term. Pick one interaction and one quadratic term that you think might be significant in predicting house prices.

Build a model using all of the original model effects and these two new terms.

1. Add all of the terms to the model.

2. Add the interaction term. Select the two terms from the **Select Columns** list and click **Cross**.

3. Add the squared term. Select the term in both the **Select Columns** list and the **Model Effects** list and click **Cross**.

Using this model, repeat the analysis illustrated in Example 1.

Questions:

a. Why did you pick the particular interaction and quadratic effect?
b. Are either of these two new terms significant?
c. Do they improve our model predictions?
d. Can you think of other predictors or terms, either in the data set or not contained in the data, that might improve the ability of our model to predict house prices?

Exercise 4.2: In this exercise we use the **BankRevenue.jmp** data.

Fit a full model to **Log(Rev_Total)** using **Log(Bal_Total)** and the other variables as model effects (using main effects only). Note, you may need to re-create these columns. Use the Minimum BIC stopping rule and stepwise regression to build your model.

  a. Compare your reduced model to that obtained using Minimum AICc in this chapter. Describe the differences in terms of the variables in the model and key statistics (adjusted R Square, RMSE, and other statistics provided).
  b. Which is the "better" model? Why? Does one model do a better job of predicting the response than the other? Explain

Exercise 4.3: Continue with the **BankRevenue.JMP** data.

Instead of fitting a model using the transformed variables, fit a model using the original (untransformed) variables. Use **Rev_Total** as the response, and **Bal_Total** and the other variables as model effects. Use stepwise and your preferred stopping rule to build the model.

  a. What are the model assumptions?
  b. Use the tools covered in this chapter to check model assumptions. Which tools should you use to check these assumptions? Explain how each tool helps check assumptions.
  c. Explain why the model assumptions are or are not met.
  d. Does it make sense to use this model to make predictions? Why or why not?

Exercise 4.4: Use the **BostonHousing.jmp** data set from the **Sample Data Directory** for this exercise. The response of interest, **mvalue**, is the median value of homes for towns in the Boston area in the 1970s.

  a. Use the tools introduced in Chapter 3 to explore the data and prepare for modeling. Are there any potential data quality issues (other than the fact that the data are from the 1970s)? Determine what actions, if any, should be taken to address data quality issues that you identify.
  b. Fit a model to **mvalue** using only **chas** and **rooms**. Recall that **rooms** is the number of rooms (rooms) and **chas** is a dummy variable (**chas**=1 indicates the town tracks the Charles River).
    i. Write down the equation for this model.
    ii. Interpret the coefficients for **chas[0]** and rooms.

    iii.   What is the predicted **mvalue** for a home that tracks the Charles River and has 6 rooms?

   c.   Fit a model to **mvalue** using all of the other variables as model effects. Use the Minimum BIC stopping rule and stepwise regression to build your model. How many terms are in the final model? Which terms are not included in the model?

   d.   Check model assumptions. Are model assumptions met? Explain.

   e.   How would a realtor, selling homes in the Boston area (in the same time period), use this model? How would a potential home buyer use this model?

References

Burnham, K. P., and D. R. Anderson. 2002. *Model Selection and Multimodel Inference, Second Edition.* Springer: New York.

Cook, R. D., and S. Weisberg. 1983. *Residuals and Influence in Regression,* New York, NY: Chapman & Hall.

Dormann, C. F., J. Elith, S. Bacher, et al. 2012. *Collinearity: A review of methods to deal with it and a simulation study evaluating their performance.* Available at http://bit.ly/1p9ml70.

Hansen, M. H., and B. Yu. 2001. "Model Selection and the Principle of Minimum Description Length." *Journal of the American Statistical Association,* Vol. 96, No. 454, Review Paper. Available at http://bit.ly/1n9f3AU; accessed 8/28/2014.

Neter, J., M. Kutner, W. Wasserman, and C. Nachtsheim. 1996. *Applied Linear Statistical Models,* 4th ed. Irwin.

Ramsey, F., and D. Shafer. *The Statistical Sleuth,* 2nd ed. Cengage Learning.

5
Logistic Regression

In the News

After spending most of 2008 predicting the success of political actors—also called politicians—it's only natural that Nate Silver (FiveThirtyEight.com) would turn his attention to the genuine article: the nominees in the major categories for the 81st Annual Academy Awards (Feb. 22 at 8 p.m. on ABC). Formally speaking, this required the use of statistical software and a process called logistic regression. Informally, it involved building a huge database of the past 30 years of Oscar history. Categories included genre, MPAA classification, the release date, opening-weekend box office (adjusted for inflation), and whether the film won any other awards. We also looked at whether being nominated in one category predicts success in another. For example, is someone more likely to win Best Actress if her film has also been nominated for Best Picture? (Yes!) But the greatest predictor (80 percent of what you need to know) is other awards earned that year, particularly from peers (the Directors Guild Awards, for instance, reliably foretells Best Picture). Genre matters a lot (the Academy has an aversion to comedy); MPAA and release date don't at all. A film's average user rating on IMDb (the Internet Movie Database) is sometimes a predictor of success; box grosses rarely are. And, as in Washington, politics matter, in ways foreseeable and not. (Nate Silver, "Oscar Predictions You Can Bet On!" in *New York Magazine*, February 15, 2009, Available at http://nymag.com/movies/features/54335/.)

Representative Business Problems

Logistic regression, among other modeling tools, can be used to model the probability that an event will occur. For example, what is the probability someone will win an Oscar, or an election, or a tennis match, based on other events that have transpired? Logistic regression can also be used to identify circumstances that make it more likely that a given event will occur.

Here are some other example uses of logistic regression:

- Identify potentially fraudulent checks or bank transactions
- Determine causes of defective items in an assembly line
- Predict flight delays
- Understand reasons behind flight risk for employees (i.e., that employees will leave and take a job elsewhere)
- Determine which content to display on a website based on mouse clicks

Preview of the End Result

Suppose a model has been developed to predict whether a customer will respond to a new catalog offer (Sharpe, 2012, p.598). The response is a yes/no categorical variable, and we want to predict the probability (*p*) that the customer will respond, based on the two predictor variables: customer age; and the number of days since the last purchase. Using methods that we will discuss in this chapter, the resulting logistic model will be in the following form:

$$log(\hat{p}/1-\hat{p}) = -3.298 + 0.08365(Age) + 0.00364(Days\ since\ last\ purchase)$$

For example, if a customer is 50 years old and there have been 100 days since the last purchase, the predicted probability of purchase is 0.78.

Looking Inside the Black Box: How the Algorithm Works

The logistic regression model allows us to fit a regression model to a set of data to develop a "typical looking" regression equation with coefficients. However, logistic regression models are different from multiple linear regression (least squares regression) models in a number of ways. First, the response is categorical, often binary (yes or no, 0 or 1). For a binary response, instead of predicting the event directly, we build a model to predict the *probability* of one level of the event (either yes or no) occurring, and that probability must be between 0 and 1. As a result, the mathematical method used to find the model coefficients is more complicated. Finally, the concept of error in a logistic model is different from the concept of what error is for a least squares regression model.

For logistic regression, the probability of an event, *p*, is related to predictive factors $(X_1, X_2, \ldots, X_k)$ by the mathematical relationship

$$log(p/(1-p)) = \beta_0 + \beta_1 X_1 + \cdots + \beta_k X_k$$

where *log(p/(1-p))* is called the *log-odds* or *logit*. The right side of this equation looks a lot like our multiple linear regression model (without the error). Rearranging this formula to solve for *p* directly, we have

$$p = 1/(1 + e^{-(\beta_0 + \beta_1 X_1 + \cdots + \beta_k X_k)})$$

In Figure 5.1, we see visual representations of two fitted logistic regression models with only one predictor, *X*. We use the "hat" notation to indicate that the parameters and predicted values for the model are estimates, based on the fitted model.

In the graph on the left, the intercept $(\hat{\beta}_0)$ of the logit is 0, and the slope $(\hat{\beta}_1)$ is 0.2. The model is written as:

$$\log(\hat{p} / (1 - \hat{p})) = \hat{\beta}_0 + \hat{\beta}_1 X = 0 + 0.2X$$

$$\hat{p} = 1 / (1 + e^{-(0 + 0.2X)})$$

At the intercept, the probability is 0.5. The slope is positive, so an increase in the value of X results in an increase in the probability of the event. In the graph on the right in Figure 5.1, the intercept for the logit is -1, and the slope is -0.5. The fitted model is written as:

$$\log(\hat{p} / (1 - \hat{p})) = -1 - 0.5X$$

$$\hat{p} = 1 / (1 + e^{1 + 0.5X})$$

For this second model, an increase in the value of X results in a decrease in the predicted probability.

Figure 5.1: The Logistic Model

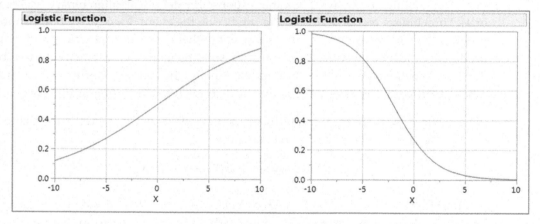

Logistic regression models are typically fit using the *method of maximum likelihood*, which is more general than the least-squares approach used in continuous response regression models. The method of maximum likelihood chooses the parameter estimates that are most consistent with the data (Sall, 2012), assuming that the logistic model is the true model that describes the data. The details for this estimation process are not shown here, but you can find further details in Neter, et al. (1996) on the estimation method.

In practice, a fitted logistic regression model can be applied in three ways. The model can be used to explain or understand how the probability of an event is influenced by the various factors. Or the model can be used to make predictions of the probability of an event. Finally, we can use the model to build a classifier rule based on the predicted probability. If we are attempting to predict the outcome of a "Yes/No" event, we could predict a future outcome based on the following rule:

$$\text{Outcome} = \begin{cases} \text{Yes} & \text{if } \hat{p} > T \\ \text{No} & \text{otherwise} \end{cases}$$

The value for T is often 0.5, but sometimes other values are used to obtain the desired level of correct classification.

Example 1:   Lost Sales Opportunities

A supplier in the automotive industry wants to increase sales and expand its market position. The sales team provides quotes to prospective customers, and orders are either won (the customer places the order) or lost (the customer does not place the order). Unfortunately, many of the quotes provided to prospective customers in the past haven't resulted in orders. To increase sales and expand, the sales team needs to understand why orders are lost. Do the historical data provide any indication why? Are there certain situations that make it more or less likely that a customer will or will not place an order?

The Data   **Lost Sales BBM.jmp**

The data set contains 550 records for quotes provided over a six-month period. The variables in the data set are:

> **Status:**   Whether the quote resulted in a subsequent order within 30 days of receiving the quote: Won = the order was placed; Lost = the order was not placed
>
> **Quote:**   The quoted price, in dollars, for the order
>
> **Time to Delivery:**   The quoted number of calendar days within which the order is to be delivered
>
> **Part Type:**   OE = original equipment; AM = aftermarket

Applying the Business Analytics Process

Define the Problem

We want to build a model that will provide insight into why some orders are won and others are lost. Understanding the factors that are related to winning or losing an order can help us identify changes that we can make in the business processes that are used to generate bids and fulfill orders to drive more accepted offers. Ultimately, this will lead to increased revenue, growth opportunities, and, hopefully, better profits. Because the response variable is a categorical binary variable (either won or lost), we will fit a logistic regression model to predict the probability that an order is won.

Prepare for Modeling

We start by getting to know our data. We graph our variables one at a time (using **Analyze > Distribution**), and then explore relationships between our response and potential predictor variables (using **Analyze > Fit Y by X**). This exploration leads to insights regarding the quality of our data, the underlying structure, and the possible need to modify or transform the data. Some of this exploration is shown here.

Our response variable, **Status,** is a two-level categorical variable. In Figure 5.2, we see that the quotes are pretty evenly split, with about 50.5% won.

Figure 5.2:   The Distribution of Status

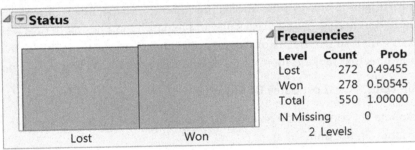

JMP plots categorical data in alpha-numeric order, so **Lost** appears first in the graph. **Lost** will also appear first in every analysis. Our future logistic model will predict the probability that an order is **Lost** rather than the probability that an order is **Won**. To change this, we set the **Value Ordering** column property for **Status** (**Column Properties > Value Ordering**, select **Won**, and click **Move Up**). After you make this change, **Won** will appear first in every graph and analysis, and our logistic model will predict the probability that an order is **Won**.

> **Note**: *If the response is coded as a 0/1 categorical variable, by default JMP will model the probability that the response is 0. To tell JMP to model probability that the response is 1, use the **Value Ordering Column Property** as described above.*

Our potential predictor variables are **Quote** (in dollars), **Time to Delivery** (the quoted delivery time in days), and **Part Type. Quote** and **Time to Delivery** are both continuous, and both distributions are right-skewed. **Part Type** is a two-level categorical variable, and there are more aftermarket quotes than original equipment. The options **Stack** and **Arrange In Rows** in the **Distribution** platform were used to create the horizontal layout in Figure 5.3.

Figure 5.3: **The Distributions of the Potential Predictors**

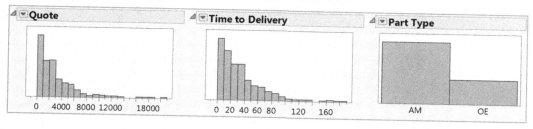

Figure 5.4: **Mosaic Plot of Status and Part Type**

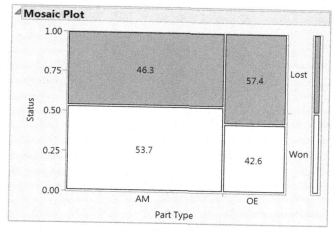

We use **Fit Y by X** to explore potential relationships between **Status** (the **Y, Response** variable) and these three factors. It appears that a higher percentage of the after market quotes are won. In this sample, 53.7% of the after market quotes were won, versus 42.6% of the original equipment quotes (Figure 5.4).

> *Note: To turn on cell labels, right-click on the graph and select **Cell Labeling > Show Percents**. To change default colors (in this case to a gray scale for printing), again right-click on the graph and use **Set Colors**.*

The **Fit Y by X** analyses of **Status** (as the **Y, Response**) versus **Quote** and **Time to Delivery** (both as **X, Factor**) results in two logistic regressions (Figure 5.5).

Figure 5.5:   Simple Logistic Model Fits of Status versus Quote and Time to Delivery

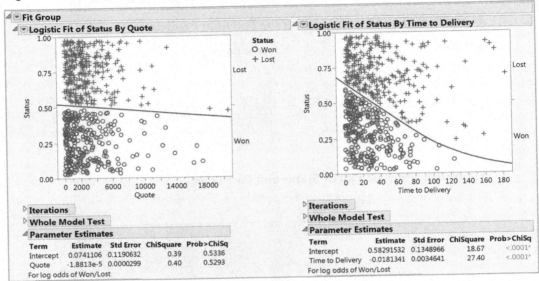

The predictor value is plotted on the *x*-axis of each graph, and the probability that an order is won is plotted on the *y*-axis. The line within the logistic graph represents the probability that an order is won as the predictor value increases. The line for **Quote** is relatively flat, indicating that the predicted probability that an order is won doesn't change much as the quoted price increases (all else held equal). The probability that an order is **Won** hovers at around 0.5, or 50%, regardless of the quoted price.

In contrast, the line in the **Time to Delivery** graph decreases as the time to deliver increases. The higher the quoted time to delivery, the lower the probability that the order is won (again, ignoring all other factors). When **Time to Delivery** is 30 days, the predicted probability that an order is won is roughly 0.5. But when **Time to Delivery** is 80 days, this drops to around 0.3.

To aid in interpretation, each point in the logistic graph is plotted at the observed value of the predictor (the *x*-axis), and on either side of the logistic line (depending on whether the order was won or lost). The points for won orders are plotted below the line, and the points for lost orders are plotted above the line. The actual position relative to the *y*-axis for each point is random, so your graphs may look slightly different from ours in Figure

5.5. Note that we have used a row legend for **Status** with colored markers to make it easier to interpret this graph (right-click on the graph and select **Row Legend**).

Build the Model

After exploring our variables and gaining an understanding of potential relationships, we fit a logistic regression model using all three predictor variables using **Analyze > Fit Model,** with **Status** as Y and the three predictors as model effects. Since our Y variable is categorical, the **Personality** automatically changes to **Nominal Logistic** (Figure 5.6).

Figure 5.6:   Fit Model Dialog Logistic

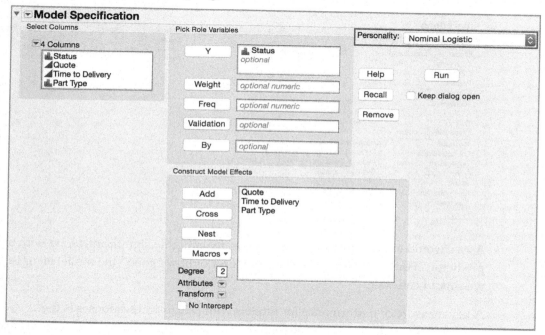

The overall model is highly significant. The *p*-value (<0.0001) is provided under **Prob>ChiSq** in the **Whole Model Test** (left, in Figure 5.7).

Of the three terms in the model, only **Time to Delivery** and **Part Type** are significant at the 0.05 level (top left, Figure 5.7).

Quote is not a significant predictor of **Status** (with a *p*-value of 0.5295), so we try fitting a simpler model without **Quote** (using the **Remove** button at the bottom of the **Effect Summary** table.) We do this because, if possible, we would like to have the simplest model that explains the variability in the data well (see the discussion of the principle of parsimony in Chapter 4).

Figure 5.7:   Logistic Model for Status with All Predictors

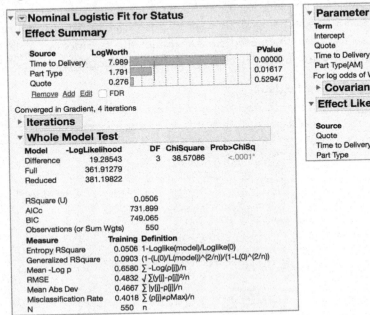

As we would expect, the resulting model is again highly significant, along with both predictors. However, there are other measures of how "good" the model is (or isn't) that we should consider.

A key measure of performance for models with categorical responses is the *misclassification rate*. In our example, the misclassification rate is 0.4018. (This is reported toward the bottom of the **Whole Model Test** section in Figure 5.8.) This means that the model-based classifications are incorrect for over 40% of the data. To put this into some perspective, if we were to categorize orders as won or lost at random (by flipping a coin, for example), the misclassification rate would be around 0.5 on average. So our misclassification rate is somewhat better than just choosing at random.

For this model, there are four possible classification outcomes:

- A won sale is correctly classified as won.

- A won sale is incorrectly classified as lost.

- A lost sale is correctly classified as lost.

- A lost sale is incorrectly classified as won.

Classifications are based on the predicted probability that an order is won. If the probability is > 0.50, then the order is classified as won. Otherwise, the order is classified as lost.

The *confusion matrix* (at the bottom of Figure 5.8) provides a breakdown of the types of correct and incorrect classifications (select **Confusion Matrix** from the red triangle). Specifically, 136 of the lost orders were classified (or predicted) as won, and 85 of the orders that were won were classified as lost.

Figure 5.8: Logistic Model for Status without Quote, with Confusion Matrix

Measure	Training	Definition
Entropy RSquare	0.0501	1-Loglike(model)/Loglike(0)
Generalized RSquare	0.0894	(1-(L(0)/L(model))^(2/n))/(1-L(0)^(2/n))
Mean -Log p	0.6584	$\sum$ -Log(p[j])/n
RMSE	0.4834	$\sqrt{\sum (y[j]-p[j])^2/n}$
Mean Abs Dev	0.4671	$\sum$ |y[j]-p[j]|/n
Misclassification Rate	0.4018	$\sum$ (p[j]≠pMax)/n
N	550	n

▶ **Lack Of Fit**

▶ **Parameter Estimates**

▼ **Effect Likelihood Ratio Tests**

Source	Nparm	DF	L-R ChiSquare	Prob>ChiSq
Time to Delivery	1	1	32.6843249	<.0001*
Part Type	1	1	5.91148329	0.0150*

▼ **Confusion Matrix**

Training

Actual Status	Predicted Won	Lost
Won	193	85
Lost	136	136

If we decide to use the model to predict the probability that an order is won, we would use the parameter estimates (Figure 5.9) to develop a prediction equation. Examining these parameter estimates allows us to understand how the predicted probability of losing an order changes as a function of **Time to Delivery** and **Part Type**. The estimate for **Time to Delivery** is negative (-0.0183), indicating that as the quoted time to deliver increases, the predicted probability that an order is won decreases. The estimate for **Part Type[AM]** is 0.2379, so the estimate for **Part Type[OE]**, which is not displayed, is -0.2379.

This indicates that the probability that an order is won is higher for aftermarket than for original equipment.

Figure 5.9: Parameter Estimates for Status

Parameter Estimates				
Term	Estimate	Std Error	ChiSquare	Prob>ChiSq
Intercept	0.48561503	0.1410173	11.86	0.0006*
Time to Delivery	-0.018344	0.0034835	27.73	<.0001*
Part Type[AM]	0.23788381	0.0982908	5.86	0.0155*
For log odds of Won/Lost				

The logistic regression model used to predict the probability an order is won is a function of these parameter estimates. We can save the probability formula to the data table using **Save Probability Formula** from the red triangle. The estimated logit for this model (Figure 5.10) is saved as a formula in a new column labeled **Lin(Won)**.

Figure 5.10: Logit for Probability of Won, Lin(Won)

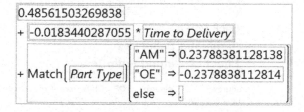

The probability of a won order, **Prob(Won)**, is then calculated from this formula (Figure 5.11).

Figure 5.11: Formula for the Probability of a Won Order, Prob(Won)

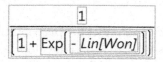

To calculate the predicted probability of a won order, for given values of our predictors, we simply plug the values into the formula for the logit, and then use the resulting value to calculate **Prob(Won)**.

For example, if the quoted delivery time is 30 days for an aftermarket order, we substitute "30" for time for delivery and use the value 0.2379 for the aftermarket effect on the logit:

$$Lin(Won) = 0.4856 - 0.0183(30) + 0.2379 = 0.1732$$

Then, we plug this value into the **Prob(Won)** formula:

$$Prob(Won) = \frac{1}{1 + e^{-Lin(Won)}} = \frac{1}{1 + e^{-0.1732}} = 0.5431$$

Of course, JMP will calculate this for us for all observations in the data table (after saving the probability formula). It will also calculate Prob(Won) and Prob(Lost) if we plug these values into a new row in the data table.

To gain a better understanding of this model, we use the *Prediction Profiler* (Figure 5.12). In Chapter 4, we built a least squares regression model and used the Prediction Profiler to explore the predicted mean of a response for different values of the predictor variables. In logistic regression, the profiler can be used in the same way to explore predicted probabilities. To access the **Prediction Profiler**, return to the **Nominal Logistic Fit window** and select **Profiler** from the red triangle.

At the average quoted time to deliver (31.79 days) the predicted probability of winning an aftermarket order is 0.535 (top, in Figure 5.12). Like the parameter estimate, the slope for time to delivery is negative, indicating that quotes with higher delivery times will have a lower probability of being won. We can also see that aftermarket quotes have a higher probability of being won than original equipment quotes. Changing the **Part Type** from **AM** to **OE**, without changing time to deliver, changes the probability of winning an order from 0.535 to 0.417 (middle, Figure 5.12).

For aftermarket products, changing the quoted time to deliver from average of 31.79 days to 10 days (bottom, Figure 5.12) changes in the predicted probability of winning an order from 0.535 to 0.632.

Figure 5.12:   Prediction Profilers for Status

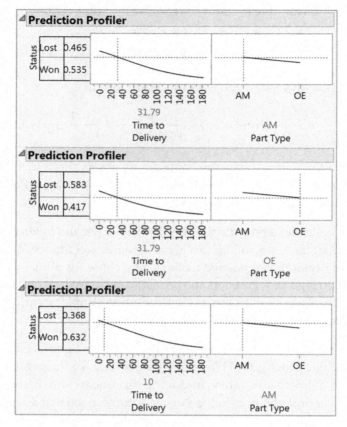

Implement the Model

In this example, the model has driven us to an interesting insight: reducing the promised delivery time could lead to an increased probability of success (i.e., winning the order). The next step is to determine how to put our model, and this new insight, to use.

One potential approach is to use the model as the basis for a new decision tool, which provides the recommended delivery time for each order. However, because of the uncertainty in predictions of this model, it may not be best to use this model to make decisions about individual sales proposals. What is clear from the model is that the time to delivery has an impact on the ability to "win" a new sale. This could help management look for ways to reduce the amount of time it takes to fill an order. This model could be used as part of the economic trade-off analysis that is needed to determine how much investment should be made in capacity increase, cycle time reduction, or other business process improvements. We explore this analysis in Exercise 5.7.

Example 2:   Titanic Passengers

The sinking of the RMS Titanic is one of the most infamous shipwrecks in history. On April 15, 1912, during her maiden voyage, the Titanic sank after colliding with an iceberg, killing 1502 out of 2224 passengers and crew. This sensational tragedy shocked the international community and led to better safety regulations for ships.

One of the reasons that the shipwreck led to such loss of life was that there were not enough lifeboats for the passengers and crew. Although there was some element of luck involved in surviving the sinking, some groups of people were more likely to survive than others. ("Titanic: Machine Learning from Disaster." From a Kaggle competition. Available at http://bit.ly/1f2crzi, data accessed 08/2014.)

The Data   Titanic Passengers BBM.jmp

This data table describes the survival status of 1309 of the 1324 individual passengers on the Titanic. Information on the 899 crew members is not included.

> **Name**:   Passenger Name
>
> **Survived**:   Yes or No
>
> **Passenger Class**:   1, 2, or 3 corresponding to 1st, 2nd, or 3rd class
>
> **Sex**:   Passenger sex
>
> **Age**:   Passenger age
>
> **Siblings and Spouses**:   The number of siblings and spouses aboard
>
> **Parents and Children**:   The number of parents and children aboard
>
> **Fare**:   The passenger fare
>
> **Port**:   Port of embarkment (C = Cherbourg; Q = Queenstown; S = Southampton)
>
> **Home/Destination**:   The home or final intended destination of the passenger

Applying the Business Analytics Process

While we don't have an intrinsic business problem, we will use this rich and storied example to dig deeper into logistic regression and explore some questions of interest about survival rates for the Titanic. For example, were there some key characteristics of the survivors? Were some passenger groups more likely to survive than others? Can we accurately predict survival? We will fit a logistic regression model using the available data to explore these questions.

We begin by examining the data, one variable at a time, two at a time, and many at a time. We only show the distribution of the response, **Survived**, and the relationship between **Survived** and the other potential predictor variables, but additional visualization and exploratory tools should also be used. Since we're interested in understanding survival rates, we've applied the **Value Ordering** column property so that **Yes** (survived) appears first in graphics and analyses, and to model the probability of survival when we build our logistic regression model.

About 38% of the passengers in our data set survived the sinking of the Titanic (Figure 5.13).

Figure 5.13:   Distribution of Survived

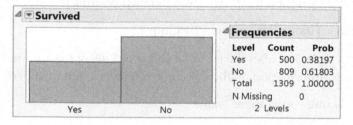

A look at two-way relationships among the response and likely predictors shows many potential (and not altogether surprising) relationships. In the examples to follow, we use **Fit Y by X** with **Survived** as **Y, Response** and the predictors as **X, Factors**.

In the mosaic plots and contingency tables in Figure 5.14, we see that first class passengers had a higher survival rate than second or third class passengers, and females fared much better than males.

Note that we have labeled the cells with the row percentages and have changed the default colors for the mosaic plots to gray scale to better see these numbers in print (right-click on the graph and select **Set Colors** and **Cell Labeling > Show Percents**.)

Figure 5.14:   Survived versus Passenger Class and Sex

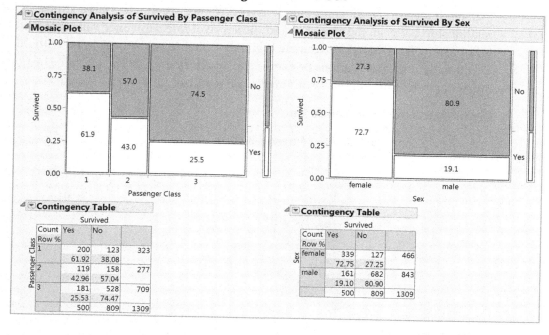

From the logistic plots in Figure 5.15, it appears that the survival rate was higher for younger passengers and for passengers traveling with more parents and children.

Figure 5.15:   Survived versus Age and Parents and Children

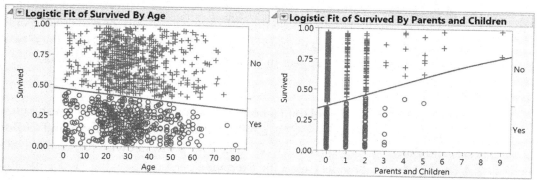

In Figure 5.15, we use **Rows > Color or Mark by Column** to mark all rows where **Survived** is **Yes** with a "+" symbol, and Survived is No with an "o" symbol. These markers appear in all future figures.

Did the passengers who paid a higher fare have a better survival rate? Looking at the logistic fit for **Survived** versus **Fare** (Figure 5.16), we see that there is a strong relationship between the fare paid and the survival rate. However, note that there are four passengers who paid extremely high fares. Should we be concerned with this? Could these four observations be exerting a large influence on the model? The potential impact of these four points is examined as an exercise.

Figure 5.16: **Survived versus Fare**

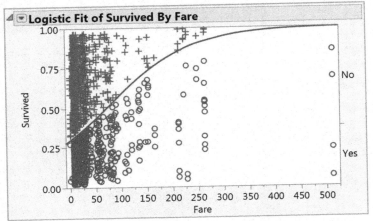

This exploratory work allows us to narrow down our list of potentially important predictor variables. We find that not all variables are important or even useful, such as **Name** and **Home/Destination**. Since we won't use these two variables in any analysis, we do a little cleanup before proceeding. We move these variables to the end of the data table, use **Column Grouping** to group them together, and then exclude these two columns so that they don't appear in variable selection lists in our analyses.

Modeling

We use **Analyze > Fit Model** to fit a nominal logistic regression model using the possible predictors **Passenger Class**, **Sex**, **Age**, **Siblings, and Spouses**, **Parents and Children**, **Fare**, and **Port**. Since we have many variables, we'll use stepwise regression to aid in the selection of the final model (select the **Stepwise** personality in the Model Specification window).

As we discussed in Chapter 4, the default stopping rule for model selection in the stepwise platform is **Minimum BIC**. Alternatively, you can use **Minimum AICc** as the stopping rule. In this situation, using either minimum AICc or BIC produces the same suggested model. But, this is not always the case.

The **Rules** option in the **Stepwise Regression Control** panel relates to how stepwise regression handles categorical variables and interaction effects. We've entered two three-level categorical variables, **Passenger Class** and **Port**. For each of these variables, two parameters can be estimated (see the **Current Estimates** section in Figure 5.17). In constructing these parameters, JMP codes the factor levels in a hierarchical fashion. (This feature is described in more detail in the *Specialized Models* JMP manual, found under the **Help > Books** menu in JMP.) The first parameter for **Port**, **Port(Q&S-C)**, combines ports Q and S into one group with port C in a separate group. The split into these two groups gives the greatest difference in the probability of survival. JMP creates a total of k-1 of these parameters for each categorical variable, where k is the number of factor levels.

To simplify the model and the interpretation of the terms in the final model, we change the rule to **Whole Effects**. With this rule, if one parameter for a variable is entered into the model, JMP will enter all of the remaining parameters for that variable into the model.

The stepwise results (after clicking **Go**), using **Minimum BIC** and **Whole Effects**, are shown in Figure 5.17. The variables selected by stepwise regression are marked with a check under **Entered** in the **Current Estimates** panel (bottom, in Figure 5.17). The five whole effects estimated by stepwise regression are **Passenger Class**, **Sex**, **Age**, **Siblings and Spouses**, and **Port**. These check boxes can be used to manually enter or remove terms from the model.

Figure 5.17:   Stepwise Logistic Regression Results

◢ ▼Stepwise Fit for Survived		

◢ **Stepwise Regression Control**

Stopping Rule: [Minimum BIC ▼] ➡ [Enter All] [Make Model]

Direction: [Forward ▼] ⬅ [Remove All] [Run Model]

Rules: [Whole Effects ▼]

[Go] [Stop] [Step]

-LogLikelihood	p	RSquare	AICc	BIC
477.43939	8	0.3228	971.018	1010.48

◢ **Current Estimates**

Lock	Entered	Parameter	Estimate	nDF	Wald/Score ChiSq	"Sig Prob"
☑	☑	Intercept[No]	1.14818061	1	0	1
☐	☑	Passenger Class{1&2-3}	0.75302892	2	57.53881	3.2e-13
☐	☑	Passenger Class{1-2}	0.56321082	2	57.53881	3.2e-13
☐	☑	Sex{female-male}	1.31631431	1	211.7167	5.8e-48
☐	☑	Age	-0.0383057	1	32.14615	1.43e-8
☐	☑	Siblings and Spouses	-0.3323162	1	10.4408	0.00123
☐	☐	Parents and Children	0	1	0.242018	0.62275
☐	☐	Fare	0	1	0.019288	0.88954
☐	☑	Port{C-S&Q}	0.53492163	2	14.2408	0.00081
☐	☑	Port{S-Q}	0.40138428	2	14.2408	0.00081

To build our logistic model, we click **Make Model** to return to the **Model Specification** window, and then click **Run**. The resulting summary of the model fit is shown in Figure 5.18. The model is highly significant, with a **Prob>ChiSq** <0.0001. However, the misclassification rate indicates that the model misclassified 21.17% of the passengers. If the goal of this modeling process was to predict survival beforehand, this would certainly be undesirable. However, our goal is to uncover the important factors related to survival. So in this analysis we are not overly concerned with the misclassification rate.

Note that **Fare** was not included in the stepwise model fit (see Figure 5.17), although by itself it was strongly related with survival. Also, **Port** was included in the model, but it is reasonable to question if that variable is truly predictive. We will explore possible reasons why predictors were or weren't included in the reduced model in an exercise.

Figure 5.18: **The Reduced Logistic Model for Survived**

Whole Model Test

Model	-LogLikelihood	DF	ChiSquare	Prob>ChiSq
Difference	228.04248	7	456.085	<.0001*
Full	477.47308			
Reduced	705.51556			

RSquare (U)	0.3232
AICc	971.085
BIC	1010.55
Observations (or Sum Wgts)	1044

Measure	Training	Definition
Entropy RSquare	0.3232	1-Loglike(model)/Loglike(0)
Generalized RSquare	0.4775	$(1-(L(0)/L(model))^{\wedge}(2/n))/(1-L(0)^{\wedge}(2/n))$
Mean -Log p	0.4573	$\sum -Log(\rho[j])/n$
RMSE	0.3821	$\sqrt{\sum(y[j]-\rho[j])^2/n}$
Mean Abs Dev	0.2929	$\sum |y[j]-\rho[j]|/n$
Misclassification Rate	0.2117	$\sum (\rho[j]\neq\rho Max)/n$
N	1044	n

Looking at the parameter estimates in Figure 5.19 allows us to gain insight into the effect of each of the predictors on the survival rate. Females, for example, have a higher survival rate than males, and young passengers have higher survival rates than older passengers.

The note at the bottom of the Parameter Estimates table, **For log odds of Yes/No**, tells us that JMP is predicting the probability that **Survival = Yes**. (Recall that we changed the value ordering at the beginning of this example.)

Figure 5.19: **Parameter Estimates for Survived**

Parameter Estimates

Term	Estimate	Std Error	ChiSquare	Prob>ChiSq
Intercept	2.28855656	0.3502206	42.70	<.0001*
Passenger Class[2-1]	-1.1270844	0.2437888	21.37	<.0001*
Passenger Class[3-2]	-0.9437064	0.2028802	21.64	<.0001*
Sex[female]	1.31659442	0.088189	222.88	<.0001*
Age	-0.0383599	0.0067077	32.70	<.0001*
Siblings and Spouses	-0.3323671	0.1030565	10.40	0.0013*
Port[C]	0.71326236	0.1882365	14.36	0.0002*
Port[Q]	-0.7578514	0.2757081	7.56	0.0060*
For log odds of Yes/No				

What do the parameter estimates actually represent? And how do we interpret the parameter estimates for **Passenger Class**, which has ordered categories and is coded as an ordinal variable?

Recall from our black box discussion at the beginning of the chapter that logistic models are based on the logit function, where $p/(1-p)$ is the *odds* of an event occurring, and the log of $p/(1-p)$ is the *logit*. From the original data we see that the probability that a second-class passenger survived was around 0.43. (See the contingency table for **Survived** versus **Passenger** class in Figure 5.14.)

The odds of surviving if you were a second-class passenger can be calculated using this ratio, $p/(1-p)$:

$$\left(\frac{\text{P(Survive|Second Class)}}{\text{P(Not Survive|Second Class)}} \right) = \left(\frac{\text{P(Survive|Second Class)}}{\text{1-P(Survive|Second Class)}} \right)$$

$$= \frac{0.43}{1 - 0.43} = 0.754$$

Similarly, we could calculate the odds for surviving for first class passengers. If we want to compare these two odds, we can calculate the *odds ratio*. If the odds ratio is close to 1, then it means that there is little difference in the two conditions with respect to helping to predict the response.

The parameter estimate for **Passenger Class[2-1]** is -1.127 (Figure 5.19). It turns out that this is an estimate of the *log of the odds ratio (or log odds)* for the odds of surviving in second class versus the odds of surviving in first class. This indicates that the overall odds of surviving in second class was $e^{-1.127} = 0.324$ lower than the odds of surviving in first class.

We see a similar result when we compare the odds of surviving for third-class versus second-class passengers. The parameter estimate for **Passenger Class[3-2]** is -0.943, indicating that the odds of surviving in third class were $e^{-0.943} = 0.389$ times lower than in second class.

To display these odds ratios, and odds ratios for all of the predictors in the model, select **Odds Ratios** from the red triangle in the **Nominal Logistic Fit** window. The odds ratios for **Passenger Class** are shown in Figure 5.20.

Figure 5.20: **Odds Ratios for Passenger Class**

Odds Ratios for Passenger Class					
Level1	**/Level2**	**Odds Ratio**	**Prob>Chisq**	**Lower 95%**	**Upper 95%**
2	1	0.3239765	<.0001*	0.1998471	0.5202023
3	1	0.126086	<.0001*	0.0783351	0.1999912
3	2	0.3891827	<.0001*	0.260867	0.5783584
1	2	3.0866439	<.0001*	1.922329	5.0038248
1	3	7.9310922	<.0001*	5.0002206	12.765669
2	3	2.5694873	<.0001*	1.7290317	3.8333716

As we saw earlier, we can use the **Prediction Profiler** (Figure 5.21) to help interpret the parameter estimates and explore the predicted survival rates for different values of the predictor variables.

Figure 5.21: **Prediction Profiler for Logistic Regression Model**

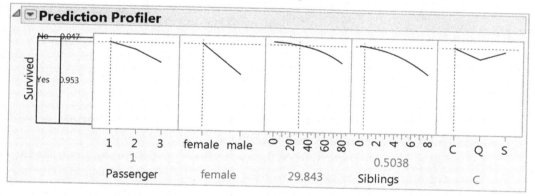

Recall that we can also save the prediction formula to the data table. The resulting logit of the probability model, saved as **Lin[Yes]** in the data table, is shown in Figure 5.22. Here, the coefficients for **Passenger Class 2** and **Passenger Class 3** are the log odds of surviving compared to **Passenger Class 1**.

Figure 5.22: **The Saved Logit Function, Lin[Yes]**

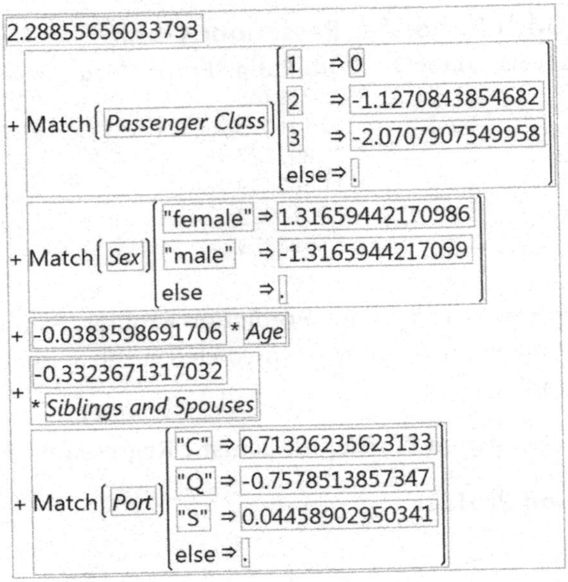

Summary

Back to our original task: Determine the characteristics of the "survivors" of the Titanic, and identify which groups of people were most likely to survive. First-class and second-class females, particularly if they were young and traveling alone, had a very good chance of surviving. The phrase "women and children first!" appeared to hold true for this group of passengers.

A final note: While this model is highly significant, could a different model provide better predictions (and a lower misclassification rate)? We've fit a model involving only main effects. What if we add interactions to the model? Would our model predictions improve? We'll explore this question in an exercise.

Key Take-Aways and Additional Considerations

- Use exploratory graphical tools and platforms, such as **Distribution**, **Fit Y by X**, **Graph Builder**, and **Scatterplot Matrix** to get to know your data, to explore potential relationships, to identify potential problems such as outliers, and to narrow down the list of potential predictor variables to include in your initial model.

- For scenarios with a large number of possible predictors, it can be determined which terms are needed in the model to make good predictions. Use stepwise regression, and explore different stopping rules to assist in selecting the best model. The use of a validation column for the stepwise procedure was not discussed in the chapter, but is covered in later chapters. This approach is recommended, wherever possible, to aid in building and assessing models.

- Misclassification is a key measure of how well the model predicts. The confusion matrix quantifies the two types of misclassification (false positives and false negatives). There are other ways to evaluate a classifier model, using ROC and Lift Curves, and these tools are discussed in later chapters in this book.

- By default, JMP will model the probability of the event that appears first in alphanumeric order. To change the target outcome category, you can use the value ordering property for the response column in the data table. You can also change the labeling of the response categories to make the labels more descriptive.

- In this chapter, both examples had two-level categorical responses. Logistic regression can also be used for multi-level (or multinomial) categorical and ordinal responses.

Exercises

Exercise 5.1: Use the **"Titanic Passengers BBM.jmp"** data set for this exercise.

- a. Re-create the analysis and graphs for Figures 5.15 and 5.16 and record the parameter estimates for each simple model.
- b. Exclude points that appear to be extreme outliers or potentially influential, and refit the simple models.
- c. Examine the parameter estimates for the models fitted with and without the extreme points. What can you conclude about the influence of these points on the models?
- d. Repeat this examination of influential points on the full, fitted logistic regression model.

Exercise 5.2: Use the **Titanic Passengers BBM.jmp** data set for this exercise.

In the one-factor-at-a-time analysis of **Survival** versus each predictive factor, **Parents & Children** appeared to be strongly related to survival. However, when the stepwise regression procedure was used to choose the best model, this factor was not included in the model.

Explore this data to try to uncover reasons for this apparent paradox. In particular, look for relationships between potential predictors to see if this can provide a reasonable explanation.

Exercise 5.3: Use the **Titanic Passengers BBM.jmp** data set for this exercise.

a. Interpret the odds ratio for **Passenger Class 1** versus **Passenger Class 3** in Figure 5.20.

b. Use the output in Figure 5.19 to calculate the odds ratio for survival for **Passenger Class 3** versus **Passenger Class 1** (show your work). Interpret this odds ratio.

Exercise 5.4: Again, use the **Titanic Passengers BBM.jmp** data set.

a. Re-create the full logistic regression model shown in Example 2.

b. Using the model parameter estimates, determine the odds ratios for not surviving for passengers in first class versus second class, second class versus third class, and female versus male.

c. Compare the estimates of the odds ratios, based on the model parameters, with the actual observed odds ratios (taking only one factor into consideration). You can determine the observed odds by doing simple percentages of "not survive" for each factor level, and calculating the odds of not surviving for each level. How closely do the odds ratios match, and if there are differences, how would you explain that?

Exercise 5.5: Use the **Equity.jmp** data from the Sample Data directory for this exercise.

This data set was first introduced in Chapter 3. Recall that the response variable is the variable **BAD**, where the value **1** indicates that the customer is a bad credit risk. If you completed the exercise in Chapter 3, you should have saved a new version of this file. We'll start with the original version, and then will return to your saved version in part c.

a. Use the **Columns Viewer**, **Distribution**, and **Graph Builder** to re-familiarize yourself with this data.

   1. Do any variables appear to be related to **BAD**? Explain.

   2. List any potential data quality issues that you observe.

b. Fit a logistic regression model for **BAD**, including all predictor variables.

   1. What is the *p*-value for the model?

2. What is the misclassification rate?

3. What are the two types of misclassification error that can occur in this example? How many misclassifications of each type were made?

4. Use the **Effect Summary** table to slowly remove non-significant terms from the model. How many terms are in your final model?

5. What is the misclassification rate for this reduced model?

6. In the context of this example, define the two types of classification error: false positive and false negative. Which type of classification error occurred more often? Explain.

7. What are the estimates (coefficients) for **DEROG** and **CLAGE**? Open the **Prediction Profiler**, and explore what happens to the predicted probability that **BAD=1** as you increase and decrease the values of these two variables.

8. You need to explain to your manager what the coefficients for **DEROG** and **CLAGE** represent. Interpret the coefficients for these two variables (in non-technical terms).

c. As we saw in Chapter 3, this data set has some data quality issues that should be addressed before modeling. Use either the **Equity.jmp** data, or use your saved version of this file (from Chapter 3) as a starting point.

1. Revisit the data preparation section in Chapter 3, and "prepare" this data for modeling. Determine how to handle missing values and other potential data quality issues (see "Common Problems with Data" in Chapter 3).

2. Fit a logistic regression model for **BAD** using this data, and again reduce this model using the **Effect Summary** table.

3. What are the terms in this reduced model?

4. How did you handle variables with missing values? Were the missing values informative? Did the missingness provide information regarding whether a customer is a good or bad credit risk? Explain.

5. What is the misclassification rate?

d. Describe differences between the final reduced models produced in parts b and c above.

1. Were any predictors included in one model but not the other?

2. Did you gain any new information or knowledge from using the "cleaned and prepared" data to build your model? Explain.

Exercise 5.6: Use the **Lost Sales BBM.jmp** data set for this exercise.

Fit a logistic regression model to predict the probability of a lost sale as a function of the quoted time to delivery and the type of part. Compare your model results with the output shown in Example 1.

Save the prediction formula for this fitted model to the data table.

Exercise 5.7 (Challenging): Use the results from Exercise 5.6 for this exercise.

a. Suppose that management would like to examine options for annual capital investment, with a goal of increasing the number of sales that would be won and also increasing the company's profitability. Because the mix of products cannot be controlled easily, adjusting the **Time to Delivery** will be used to see if this can lead to improved order success rate and profitability. The following table contains engineering-based and financial-based estimates to achieve different levels of **Time to Delivery**. For each level of investment and expected improvement in **Time to Delivery**, calculate the probability of winning a sale for AM and OE parts.

Annualized Investment Cost ($M)	Expected Time to Delivery (days)	P[Won Sale] for AM Parts	P[Won Sale] for OE Parts
0 (baseline)	32		
1	30		
2	25		
3	23		
6	22		
10	20		

b. The company expects to see 1000 requests for quotes for AM parts and 500 requests for OE parts annually. The average profit for an AM parts order is $50,000, while the average profit for an OE parts order is $75,000. For each row in the table that you created, calculate the expected annual profit. (For a particular part type and time to delivery, the expected annual profit is [(Number of Quotes)*P[Won Order]*Expected Profit of a Won order)]).

c. Based on this analysis, what level of investment would be recommended to increase profitability? (Hint: Compare the expected annual profit with the annualized investment cost.)

d. Bonus: The analysis above assumes that all the values used in the economic analysis are fixed. What if there is uncertainty associated with the assumed values for numbers of orders per year and revenue per order? How would that impact your analysis and recommendations? (Hint: You can use the simulator in the JMP prediction profiler to help answer this question.)

References

Neter, J., M. Kutner, W. Wasserman, and, C. Nachtsheim. 1996. *Applied Linear Statistical Model*, 4th ed. Irwin.

Sall, J., A. Lehman, M. Stephens, and L. Creighton. 2012. JMP *Start Statistics*, 5th ed. SAS Institute Inc.

Sharpe, Norean D., Richard D. De Veaux, and Paul F. Velleman. 2011. *Business Statistics*, 2nd ed. Boston, MA: Pearson Education, Inc.

Decision Trees

In the News

> Retail analytics firm Dunnhumby Ltd. hosted a hackathon Saturday in
> Cambridge Massachusetts through Kaggle Inc.to predict how future consumer
> goods will sell. Kaggle is a company that hosts online predictive modeling
> competitions. By weekend's end, not only did Dunnhumby gain a predictive
> model that it could eventually offer to its retail client-base, but also several hiring
> prospects discovered at the event.
>
> With thousands of stores vying for shares of consumers' wallets, the success or
> failure of a new product often hinges on the first few weeks of sales. The earlier a
> retailer knows how well a product is resonating with consumers, the sooner it
> can decide whether to offer it in more stores or abandon it. Dunnhumby, which
> builds algorithms to help retailers better target their customers, may use the
> winners' predictive model to package data that it sells to its retail customers.
> (Clint Boulton, "Hackathon Helps Retail Analytics Firm 'Find Talent, Learn New
> Ideas'," *CIO Journal*, May 13, 2013. Available at http://on.wsj.com/1tF6Nka, ac-
> cessed 10/2014.)

A variety of powerful predictive modeling tools were used by teams competing in the
Dunhumby hackathon. A core tool in the analytics toolbox, and the focus of this chapter,
is the decision tree.

Representative Business Problems

Decision trees fall into two general categories, classification trees and regression trees. If
the target variable is categorical (nominal or ordinal), then a classification tree is used to
predict the probability of a particular outcome based on a set of predictors. If the target
variable is continuous, then a regression tree is used to predict the mean of the response.
For example, if a charitable foundation is studying potential donors, then a classification
tree could be used to predict *whether a person will contribute*, while a regression tree could
be used to predict *how much* a person might contribute.

Decision trees are used for both exploratory data analysis and for predictive modeling.
They are relatively easy to understand and explain, particularly to a non-technical
audience.

Here are some typical questions that can be addressed with decision tree models:

- Will a patient be readmitted to the hospital over the next 30 days?
- Will a new movie be a blockbuster? How much will it earn?
- Will a given candidate win an election?
- What is the probability that a flight will be delayed?
- Which variables are most important in predicting housing prices?
- What is the risk that an applicant will default on a loan?
- What factors might be causing low yields or manufacturing defects?

Preview of the End Result

A decision tree consists of a set of conditional rules, based on simple decision thresholds. The "tree" is essentially a series of nested if-then statements that lead to a classification or prediction.

For example, conditional rules that could be used in a credit risk assessment might look like this:

- If income is > 100,000 and credit rating is > 600, then credit risk is low; but if credit rating is < 600, then credit risk is high, and
- If income is <=100,000 and credit rating is < 700, then credit risk is high; but if credit rating is > 700, then credit risk is low.

We can visualize this set of rules in a hierarchical diagram that looks like an upside down tree (Figure 6.1).

Figure 6.1: Example Decision Tree

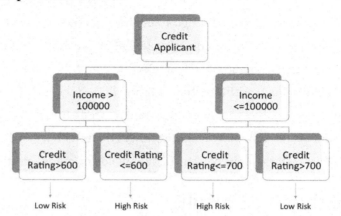

Looking Inside the Black Box: How the Algorithm Works

To understand the algorithm used to build decision trees in JMP, consider the data in **Lost Sales BBM.jmp**. The data are described in Chapter 5, "Logistic Regression." The response of interest is the categorical variable **Status** (Won or Lost), and there are three potential predictors: **Quote**, **Time to Delivery**, and **Part Type**. **Quote** and **Time to Delivery** are continuous, and **Part Type** is nominal.

The algorithm used in building a decision tree depends on the type of response. When the response is categorical, the model that we build is known as a *classification tree*. If the response is continuous, then the model that we build is a *regression tree*. We can use both continuous and categorical predictors for either type of model. In this example we'll build a classification tree. Later in the chapter, we'll see two detailed examples, the first involving a classification tree and the second using a regression tree.

Note that in the section, "Other General Modeling Considerations," we use the **Lost Sales** example to introduce some general modeling considerations that apply to the various modeling approaches covered in this book.

Classification Tree for Status

To build a decision tree to classify **Status**, we start by looking at subsets of the data based on the values of each of the predictors. These subsets form the "branches" of the tree (as shown in Figure 6.1). But how does that process begin? We now look at the logic for forming the tree and selecting each of the branches in the tree.

For instance, suppose we consider these two subsets, **Part Type=OE** and **Part Type=AM**. Notice that these two subsets, when put together, make up the entire data set. We call this type of sub-setting a "split" because it splits the data into two parts. The question is does this split cause the two subsets of the data to be more similar within each subset but less similar with regard to **Status**? If so, then this is a good candidate split to use to make a prediction about **Status.**

The overall proportion of **Status=Won** is 278/550 = 50.54%, and the proportion of **Status=Lost** is 272/550 = 49.45%. (This is shown in Figure 6.2, and in the initial **Partition** graph in Figure 6.6.)

Figure 6.2:   Lost Sales, Distribution of Status

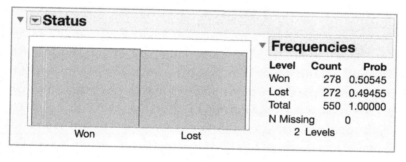

In this example, we have three potential predictors, **Quote**, **Time to Delivery**, and **Part Type**. As we form our tree, we have two decisions to make: *Which variable* to split on and *what value* of this variable to split at.

To illustrate how the underlying algorithm evaluates different splits, we use the **Analyze > Fit Y by X** platform. First, we split the data by the variable **Part Type** (use **Status** as **Y, Response** and **Part Type** as **X, Factor**). This results in the mosaic plot and contingency table in Figure 6.3. (Note that we have used the red triangle for the Contingency table to turn off some default options.)

This results in the following proportions: 53.67% **Won** and 46.33% **Lost** for the **AM** group and 42.58% **Won** and 57.42% **Lost** for the **OE** group. There is a higher proportion of **Won** orders in the **AM** group than there is in the **OE** group.

Figure 6.3:   Lost Sales, Contingency of Status by Part Type

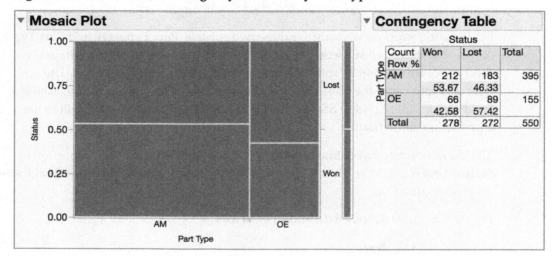

Now consider the continuous variable **Time To Delivery**. Since this variable can take on many values (it ranges from 0 to 183), there are many potential split points. In Figure 6.4, we have manually split the data into two buckets, <= 50 days and > 50 days. (We created a new column, and used the **Formula Editor** as shown in Chapter 3 to bucket the data.) This split appears to drive a separation in the proportion of **Won** and **Lost** sales. But this is just one of many split points that we can consider. How do we determine the split that drives the greatest separation?

Figure 6.4:   Lost Sales, Contingency of Status by Delivery Split

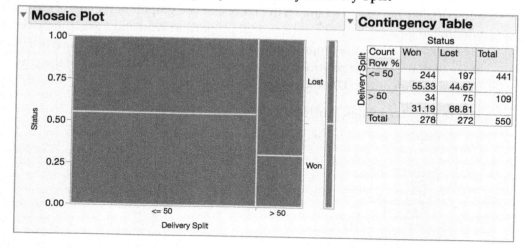

Statistical Details Behind Classification Trees

A measure of the dissimilarity in the proportions between the two split groups is the *likelihood ratio chi-square statistic* and its associated *p*-value. The lower the *p*-value, the bigger the difference between the groups. When JMP calculates this chi-square statistic in the **Partition** platform, it is labeled G^2; and the *p*-value that is calculated is adjusted to account for the number of splits that are being considered. The adjusted *p*-value is transformed to a log scale using the formula **-log10(adjusted p-value)**. This value is called the *LogWorth*. The bigger the LogWorth value, the better the split is (Sall, 2002). This statistic is used to determine the best split location and split value.

> *Note: For regression trees, which have a continuous response, SS (Sum of Squares) is reported instead of G^2. The construction of regression trees is analogous to what we describe for classification trees, but splits are based dissimilarity of the mean of the response in the split groups rather than the dissimilarity in the proportions. For regression trees, the LogWorth is also used to determine the optimal split.*

To find the split group that has the largest difference between groups (and the corresponding largest value of LogWorth), we need to consider all possible splits. For each variable, the best split location is determined, and the split with the highest LogWorth is chosen as the optimal split location.

To illustrate, we launch the **Partition** platform (use **Analyze > Modeling > Partition**), and populate the dialog as shown in Figure 6.5. The graphic at the top of the analysis window in Figure 6.6 displays the overall proportion **Won** and **Lost**. The *y*-axis value for the horizontal line is the proportion **Won**. JMP reports the G^2 and LogWorth values, along with the best *cut points* for each variable, under **Candidates** (use the gray disclosure icon next to **Candidates** to display).

Figure 6.5:   Lost Sales, Partition Dialog

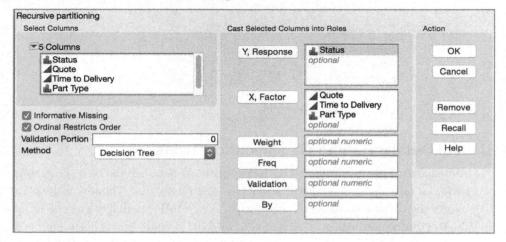

Figure 6.6: Lost Sales, First Candidate Split

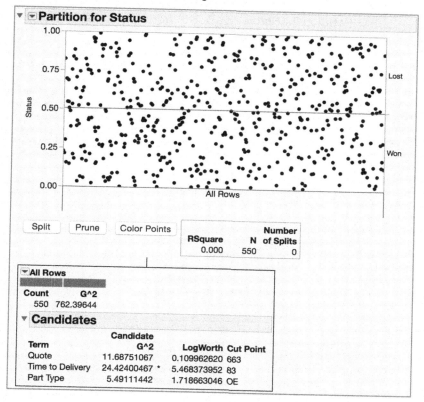

In this example, the term with the highest LogWorth (5.468) is **Time to Delivery**, and the cut point (or split value) is 83. The resulting tree after the first split is shown in Figure 6.7 (click **Split** below the graph). The number of observations in each subset and the split probabilities are shown for each branch. To display these values, use the **Show Split Count** option under the top red triangle, **Display Options**.

The **Partition** graph (top in Figure 6.7) updates to display the counts in each subset (as the area on either side of the vertical split line and the corresponding points) and the split probabilities (the values on the *y*-axis corresponding to the horizontal lines).

Figure 6.7: **Lost Sales, Status after First Split**

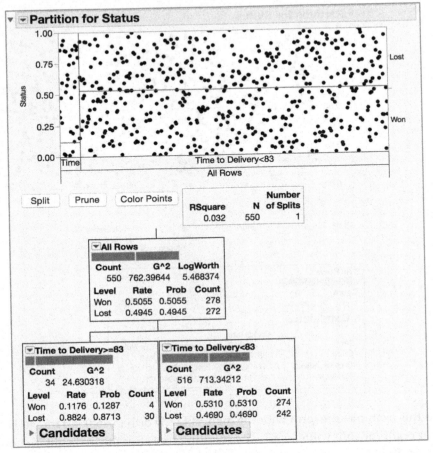

The new candidate splits within each branch, or *node*, are shown in Figure 6.8.

The split with the highest LogWorth across all of the nodes is the next optimal split location. Although **Quote** in the **Time to Delivery < 83** branch has the highest G^2 (13.723), **Part Type**, in the same branch, has the highest LogWorth (1.97). Since the rule is to always chose the split with the highest LogWorth value, the next split is on **Part Type** in the **Time to Delivery < 83** branch.

Figure 6.8: **Lost Sales, Candidate Split for Status after First Split**

Time to Delivery>=83			
Count	**G^2**		
34	24.630318		
Level	**Rate**	**Prob**	**Count**
Won	0.1176	0.1287	4
Lost	0.8824	0.8713	30

▼ **Candidates**

	Candidate		
Term	**G^2**	**LogWorth**	**Cut Point**
Quote	4.614220944 *	0.5319112211	1753
Time to Delivery	2.646823915	0.2297340156	90
Part Type	0.005382993	0.0261738728	AM

Time to Delivery<83			
Count	**G^2**		
516	713.34212		
Level	**Rate**	**Prob**	**Count**
Won	0.5310	0.5310	274
Lost	0.4690	0.4690	242

▼ **Candidates**

	Candidate		
Term	**G^2**	**LogWorth**	**Cut Point**
Quote	13.72304776 <	0.350819593	663
Time to Delivery	10.87784087	1.693560594	30
Part Type	6.51065624 >	1.969683983	OE

The tree after five splits is displayed in Figure 6.9 (the split probabilities and counts are not shown).

Figure 6.9: **Lost Sales, Partition Tree for Status after Five Splits**

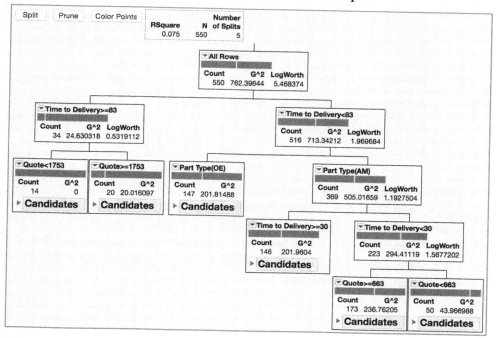

The algorithm has produced a tree that can be used for making predictions. For each node in the tree, we have an estimate of the probability that **Status** is **Won** or **Lost**. These probabilities are summarized in the **Leaf Report** (shown in Figure 6.10). The **Leaf**

Report is an option under the top red triangle. The highest probability that **Status** is **Won** (0.8338), shown in the last row of the leaf report, has four splits: two splits on **Time to Delivery** (at 83 and at 30), one split on **Part Type,** and one split on **Quote** (at 663). Here is the interpretation of this leaf, or decision rule: When **Time to Delivery** < 83, **Part Type** = AM, **Time to Delivery** < 30, and **Quote** < 663, the predicted probability that **Status** = **Won** is 0.8338 (and the probability that **Status** = **Lost** is 0.1662). Since two splits are on **Time to Delivery**, this decision rule can be simplified: When **Time to Delivery** < 30, **Part Type** = AM, and **Quote** < 663, the predicted probability that **Status** = **Won** is 0.8338.

Figure 6.10:   Lost Sales, Leaf Report after Five Splits

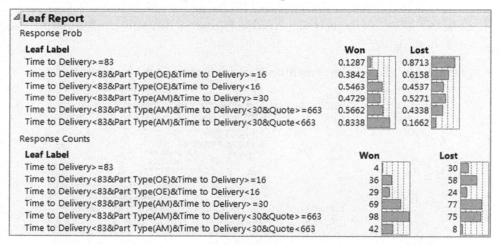

The probability formula can be viewed by saving the prediction formula to the data table (select **Save Columns > Save Prediction Formula** from the top red triangle). The prediction formulas for both the probability of **Status** = **Won** and **Status** = **Lost** are saved in new columns in the data table. In this example, these columns are labeled **Prob(Status==Won)** and **Prob(Status==Lost)**. The formula for the probability that **Status** = **Won** is shown in Figure 6.11.

Figure 6.11: Lost Sales, Probability Formula for Won after Five Splits

```
If  ! Is Missing( Time to Delivery )
    & Time to Delivery >= 83           ⇒ If  Is Missing( Quote ) | Quote < 1753  ⇒ 0.03118545454545
                                             else                                ⇒ 0.21275151515152
    else                               ⇒ If  Part Type == "OE"  ⇒ 0.44262165488858
                                             Part Type == "AM"  ⇒ If  ! Is Missing( Time to Delivery )
                                                                      & Time to Delivery >= 30         ⇒ 0.47288318514472
                                                                      else                             ⇒ If  ! Is Missing( Quote ) & Quote >= 663  ⇒ 0.56623663667955
                                                                                                             else                                  ⇒ 0.83382695651456
                                             else               ⇒ 0.53095832600668
```

A most likely classification column is also created when the prediction formula is saved to the data table (stored as **Most Likely Status**). By default, the most likely classification for a row is the outcome with the highest predicted probability. In this example, since we have two possible outcomes, this equates to using a cutoff for the probability of 0.5.

How good is this model at classifying sales opportunities? The overall classification accuracy is reported as the **Misclassification Rate** in the **Fit Details** report, and the details are provided in the **Confusion Matrix** (see Figure 6.12). With a misclassification rate of 0.3927, our model clearly doesn't do a very good job at the cutoff of 0.5. Note that changing this cutoff can, in some cases, improve the misclassification rate.

Figure 6.12: Lost Sales, Misclassification Rate, and Confusion Matrix

Fit Details

Measure	Training	Definition
Entropy RSquare	0.0746	1-Loglike(model)/Loglike(0)
Generalized RSquare	0.1311	(1-(L(0)/L(model))^(2/n))/(1-L(0)^(2/n))
Mean -Log p	0.6414	∑ -Log(p[j])/n
RMSE	0.4757	√ ∑(y[j]-p[j])²/n
Mean Abs Dev	0.4536	∑ |y[j]-p[j]|/n
Misclassification Rate	0.3927	∑ (p[j]≠pMax)/n
N	550	n

Confusion Matrix

Training

Actual	Predicted	
Status	Won	Lost
Won	169	109
Lost	107	165

We provide additional information on misclassification rates and interpreting the confusion matrix in the case studies later in this chapter.

Other General Modeling Considerations

Exploratory Modeling versus Predictive Modeling

The types of classification and regression tree models that can be fit with the **JMP Pro Partition** platform include **Decision Trees** (the focus of this chapter), **Bootstrap Forests**, and **Boosted Trees** (the latter two are covered in Chapter 9). These models can be used for both exploratory modeling and predictive modeling.

In *exploratory modeling*, the emphasis is on finding factors in the model that have the strongest relationship with the response, but using the model to make predictions about future events is not usually of direct interest. Exploratory modeling will often lead to further analysis of important variables using additional modeling methods, or to further study of potential cause-and-effect relationships using formally designed experiments.

One measure of how much a variable contributes to the model is the number of times the model splits on that variable. A second measure is the accumulated split statistic, G^2. Using the **Column Contributions** option in the **Partition** report window, we can see which factor columns contributed the most to the predictive model (select **Column Contributions** from the top red triangle). The column contributions table for the model above is shown in Figure 6.13.

Figure 6.13:   Lost Sales, Column Contributions

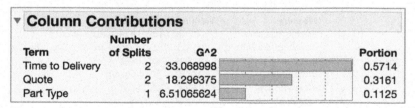

▼ Column Contributions				
Term	Number of Splits	G^2		Portion
Time to Delivery	2	33.068998		0.5714
Quote	2	18.296375		0.3161
Part Type	1	6.51065624		0.1125

In *predictive modeling*, the emphasis is on predicting the response variable as accurately as we can. Understanding which factors are most important and how they are related to the response is generally of less importance. In predictive modeling, emphasis is put on selecting a model that is neither *under-fit* (too simple) nor *over-fit* (too complex). The decision tree modeling algorithm makes it very easy to build over-fit models. If you imagine that you continued to find splits in the data, with no restrictions, you could build a decision tree where each final node in the tree contains only one row, resulting in a highly over-fit model.

One simple approach to preventing this from occurring is to specify the minimum number of rows that can be in any final tree node. The **Minimum Size Split** setting in the decision tree platform has a default of 5. But, it can be set to any integer value (number of rows) or any decimal value between 0 and 1 (proportion of rows).

Another approach to prevent over-fitting is to use *model cross-validation*. We cover this topic in depth in Chapter 8. However, because model over-fitting is such a critical issue in predictive modeling, and decision trees are especially prone to over-fitting, we briefly introduce the concept here.

Model Cross-Validation

When we build a predictive model, there is a risk that we model the noise in the particular data set, rather than modeling the true relationship between the predictors and the response. The resulting model may be overly complicated (*over-fit*), or may not perform well when applied to new data. Cross-validation helps us guard against this.

Here's the basic idea (again, we'll cover this much more formally in Chapter 8). When we use model cross-validation, we hold a subset of our data out of the modeling process. The portion of data used to build the model is often referred to as the *training set*, and the data held out is often referred to as the *validation set* (or the *hold out set*). We build the model using the training data, and then apply the model to the validation data and see how well it performs. In the **Partition** platform, the validation set tells JMP when to stop splitting.

To briefly illustrate using the **Lost Sales** data, we return to the **Partition** dialog. There are a number of ways to use cross-validation in JMP Pro. A validation option available in both JMP and JMP Pro is the **Validation Portion**. Here, we tell JMP to hold out 30% of the data using the **Validation Portion** field (see Figure 6.14).

Figure 6.14: Lost Sales, Partition Dialog with Holdout

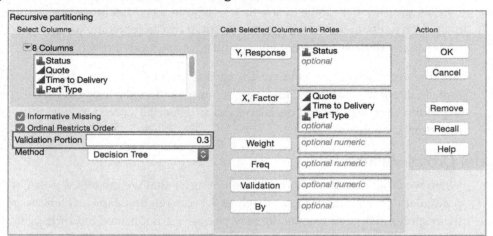

Note that the validation portion is randomly chosen, so you may see different results if you're following along. To obtain the results shown in Figure 6.16, set the random seed to 123 before launching the **Partition** platform. The random seed can be set with the **Random Seed Reset** add-in, which can be installed from the **JMP File Exchange** (https://community.jmp.com/docs/DOC-6601).

As the model is built, statistics for both the training set and the validation set are captured (Figure 6.15). As before, we can build the model one split at a time (using the **Split** button). However, since we have a validation set, we can click **Go** to automatically build the model. When this option is used, the final model will be based on the model with the *maximum value of the Validation RSquare* statistic.

Figure 6.15: **Lost Sales, Partition Dialog with Holdout**

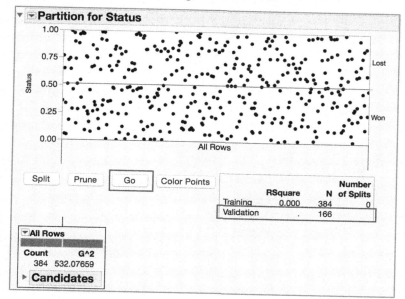

The final model has seven splits (shown at the top of the output in Figure 6.16). The **Split History** report (Figure 6.17) shows how the RSquare value changes for training and validation data after each split. The vertical line is drawn at the number of splits used in the final model.

Figure 6.16:   Lost Sales, Model with Validation Portion

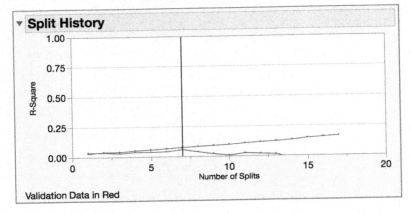

Figure 6.17:   Lost Sales, Model Spit History

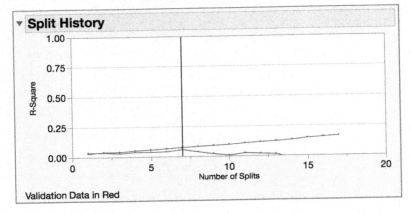

This illustrates both the concept of over-fitting and the importance of cross-validation. With each split, the RSquare for the training data continues to increase. However, after seven splits, the validation RSquare (the lower line in Figure 6.17) starts to decrease. For the validation set, which was not used to build the model, additional splits are not improving our ability to predict the response.

Dealing with Missing Values

Although we don't face this problem in this example, missing values for predictor variables is a common issue in most modeling situations. The default approach in the JMP Partition platform is to consider that *the fact that a value is missing* might actually be informative. For example, if a factor is a survey response question about a political issue, a non-response may be coded as a missing value, but that non-response may actually have some relationship to the response of interest. This approach to handling missing values is referred to in JMP as **Informative Missing**.

When this option is used, rows for continuous factors that have missing values are first considered to be on the low end of the split for each possible split evaluated. Then they are considered to be on the high end of the split for each possible split. The split with the best LogWorth is selected. Once the "missing" rows have been put into the high or low side, they remain there for further splits on that factor.

For categorical factors, a missing value code is used to indicate a missing value, and this missing value code is treated as another level of the categorical factor. If informative missing is not used, then an imputation approach is used, where rows with missing factor values are randomly assigned to either one side of the split or the other.

Decision Tree Modeling with Ordinal Predictors

For factors that are Ordinal (ordered categories), it is often desired to preserve the order of the categories in the modeling process. The default option in JMP is **Ordinal Restricts Order**. This preserves ordering so that splits on Ordinal factors will have levels that are next to each other. For instance, if there is an Ordinal factor **Rating** that can take on ordered values 1, 2, 3, 4, and 5, preserving the order would mean that the only possible initial splits on this factor are (1 versus 2,3,4,5), (1,2 versus 3,4,5), (1,2,3 versus 4,5) and (1,2,3,4 versus 5).   This option can be deselected, however, which would allow for unordered splits for this factor, such as 1,3 versus 2,4,5.

Example 1:   Credit Card Marketing

A bank would like to understand the demographics and other characteristics associated with whether a customer accepts a credit card offer. Observational data is somewhat limited for this kind of problem, in that often the company sees only those who respond to an offer. To get around this, the bank designs a focused marketing study, with 18,000 current bank customers. This focused approach allows the bank to know who does and does not respond to the offer, and to use existing demographic data that is already available on each customer.

The designed approach also allows the bank to control for other potentially important factors so that the offer combination isn't confused or confounded with the demographic factors. Because of the size of the data and the possibility that there are complex relationships between the response and the studied factors, a decision tree is used to find out if there is a smaller subset of factors that may be more important and that warrant further analysis and study.

The Study

Each customer receives an offer for a new credit card that has a particular type of reward program associated with it. The type of mailing is varied, with some customers receiving a letter and others receiving postcards. The reward type and mailing type are varied in a balanced random fashion across the prospective customer demographics. After a set period of time, whether the individual has responded positively to the mailer and has opened a new credit card account is recorded.

The Data   Credit Card Marketing BBM.jmp

The data set consists of information on the 18,000 current bank customers in the study. The variables are:

>**Customer Number:** A sequential number assigned to the customers (this column is hidden and excluded – this unique identifier will not be used directly)
>
>**Offer Accepted:** Did the customer accept (Yes) or reject (No) the offer
>
>**Reward:** The type of reward program offered for the card
>
>**Mailer Type:** Letter or postcard
>
>**Income Level:** Low, Medium, or High

 # Bank Accounts Open: How many non-credit card accounts are held by the customer

Overdraft Protection: Does the customer have overdraft protection on their checking account(s) (Yes or No)

Credit Rating: Low, Medium, or High

 # Credit Cards Held: The number of credit cards held at the bank

 # Homes Owned: The number of homes owned by the customer

Household Size: Number of individuals in the family

Own Your Home: Does the customer own their home (Yes or No)

Average Balance: Average account balance (across all accounts over time)

Q1 Balance: Average balance for Q1 in the last year

Q2 Balance: Average balance for Q2 in the last year

Q3 Balance: Average balance for Q3 in the last year

Q4 Balance: Average balance for Q4 in the last year

Applying the Business Analytics Process

Define the Problem

We want to build a model that will provide insight into why some bank customers accept credit card offers. Because the response is categorical (either Yes or No) and we have a large number of potential predictor variables, we use the **Partition** platform to build a classification tree for **Offer Accepted**.

Prepare for Modeling

As always, we start by getting to know our data. Since we have a relatively large data set with many potential predictors, we use **Cols > Columns Viewer** to create numerical summaries of each of our variables (see Figure 6.18).

Figure 6.18:   Credit, Summary Statistics for All Variables with Columns Viewer

Credit Card Marketing BBM with Scripts.jmp (18000 rows, 17 columns)

▸ **Columns View Selector**

▾ ☐ **Summary Statistics**

16 Columns [Clear Select] [Distribution]

Columns	N	N Missing	N Categories	Min	Max	Mean	Std Dev
Offer Accepted	18000	0	2	.	.	.	.
Reward	18000	0	3	.	.	.	.
Mailer Type	18000	0	2	.	.	.	.
Income Level	18000	0	3	.	.	.	.
# Bank Accounts Open	18000	0	.	1	3	1.2557777778	0.4725005801
Overdraft Protection	18000	0	2	.	.	.	.
Credit Rating	18000	0	3	.	.	.	.
# Credit Cards Held	18000	0	.	1	4	1.9035	0.7970088081
# Homes Owned	18000	0	.	1	3	1.2034444444	0.4273412028
Household Size	18000	0	.	1	9	3.4990555556	1.1141819457
Own Your Home	18000	0	2	.	.	.	.
Average Balance	17976	24	.	48.25	3366.25	940.51556242	350.29783672
Q1 Balance	17976	24	.	0	3450	910.45065643	620.07706023
Q2 Balance	17976	24	.	0	3421	999.39218959	457.402268
Q3 Balance	17976	24	.	0	3823	1042.0336004	553.45259941
Q4 Balance	17976	24	.	0	4215	810.18580329	559.00136526

Under **N Categories**, we see that each of our categorical variables has either two or three levels. **N Missing** indicates that we are missing 24 observations for each of the balance columns. (Further investigation indicates that these values are missing from the same 24 customers.) The other statistics provide an idea of the centering, spread, and shapes of the continuous distributions.

Next, we graph our variables one at a time using **Analyze > Distribution**. In Figure 6.19, we see that only around 5.68% of the 18,000 offers were accepted.

Figure 6.19:   Credit, Distribution of Offer Accepted

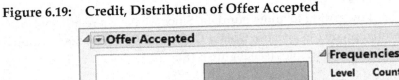

Level	Count	Prob
Yes	1023	0.05683
No	16977	0.94317
Total	18000	1.00000

N Missing   0

2 Levels

We select the **Yes** level in **Offer Accepted** and then examine the distribution of accepted offers (the shaded area) across the other variables in our data set (the first 10 variables are shown in Figure 6.20).

Our two experimental variables are **Reward** and **Mailer Type**. **Points** and **Air Miles** have more of the offers accepted than **Cash Back**, while **Postcard** has more of the accepted offers than **Letter**. Offers also appear to be accepted at a higher rate by customers with low to medium income, no overdraft protection, and low credit rating.

Note that both **Credit Rating** and **Income Level** are coded as **Low**, **Medium**, and **High**. Peeking at the data table, we see that the modeling types for these variables are both *nominal*, but they should be coded as *ordinal* variables. To change the modeling types, right-click on the modeling type icon in the data table or in any dialog window, and then select the correct modeling type.

Finally, we explore relationships between our response and potential predictor variables using **Analyze > Fit Y by X**. This two-way analysis shows potential relationships between **Offer Accepted** and several variables, including **Reward**, **Mailer Type**, **Income Level**, and **Credit Rating.**

The first two analyses are shown in Figure 6.21. Note that tests for association between the predictors and **Offer Accepted** are also provided by default (these are not shown in Figure 6.21), and additional statistical options and tests are provided under the top red triangles.

Figure 6.20: Credit, Distribution of First 10 Variables

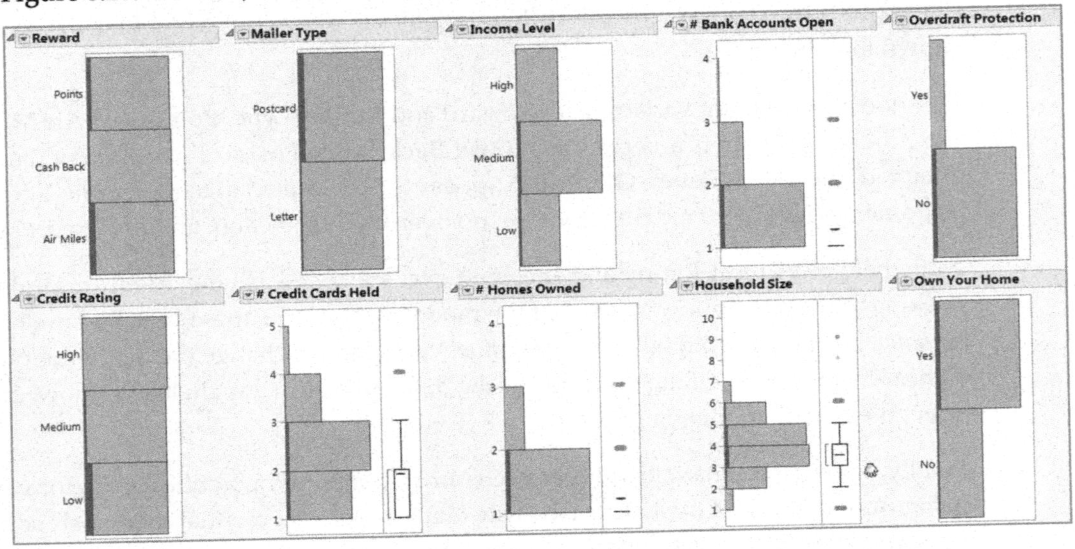

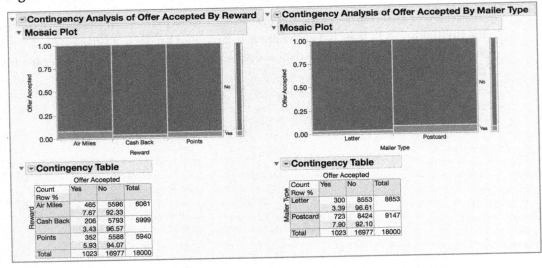

Figure 6.21: Credit Fit Y by X, Offer Accepted versus Reward and Mailer Type

Although we haven't thoroughly explored this data, here's what we have learned thus far:

- Only a small percentage, roughly 5.68%, of the offers are accepted.

- We are missing some data for the **Balance** columns, but are not missing values for any other variable.

- Both of our experimental variables (**Reward** and **Mailer Type**) appear to be related to whether or not an offer is accepted.

- Two variables, **Income Level** and **Credit Rating**, should be coded as **Ordinal** instead of **Nominal**.

Again, we encourage you to thoroughly explore your data, and to investigate and resolve potential data quality issues before building any model. While we have only superficially explored the data in this example (and in future examples), this exploration is primarily from the perspective of getting to know our variables used in the case study and is intentionally brief.

Build the Model

Having a good sense of data quality and potential relationships, we now fit a partition model to the data using **Analyze > Modeling > Partition**, with **Offer Accepted** as **Y, Response** and all of the other variables as **X, Factor**. The resulting dialog window is displayed in Figure 6.22.

There are two methods for model cross-validation available from the **Partition** dialog window in JMP Pro: Specify a **Validation Portion** or select a **Validation** column. This study is exploratory in nature. However, since we have a relatively large data set, we'll use a hold out portion (30%) to protect against over-fitting. To obtain the same results, again use the **Random Seed Reset** add-in to set the random seed to 123 (before launching the **Partition** platform).

Two additional options in the dialog window are **Informative Missing** and **Ordinal Restricts Order**, which were described earlier in this chapter. These are selected by default. In this example, we have two ordinal predictors, **Credit Rating** and **Income Level**. We also have missing values (for the five balance columns).

Figure 6.22:   Credit, Partition Dialog Window

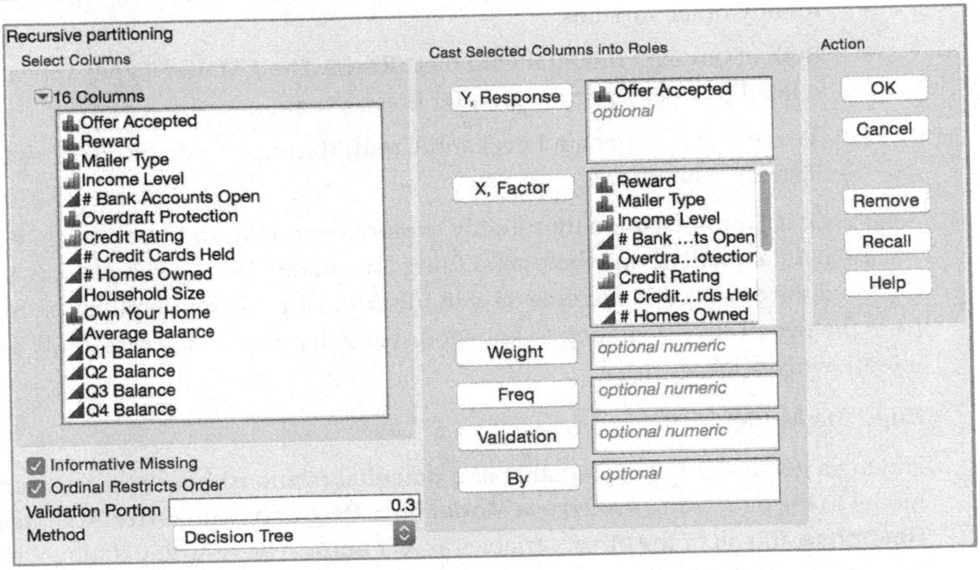

The initial results show the overall breakdown of **Offer Accepted** (Figure 6.23). Note that we have changed some of the default settings:

- To remove the points from the graph (since we have a relatively large data set), click on the **top red triangle** and select **Display Options > Show Points**.

- To show response rates in the tree nodes, select **Display Options > Show Split Count** from the top red triangle.

A peek at the candidates indicates that the first split will be on **Credit Rating**, with **Low** on one node and **High** and **Medium** on the other (Figure 6.24).

Figure 6.23: Credit, Partition Initial Window

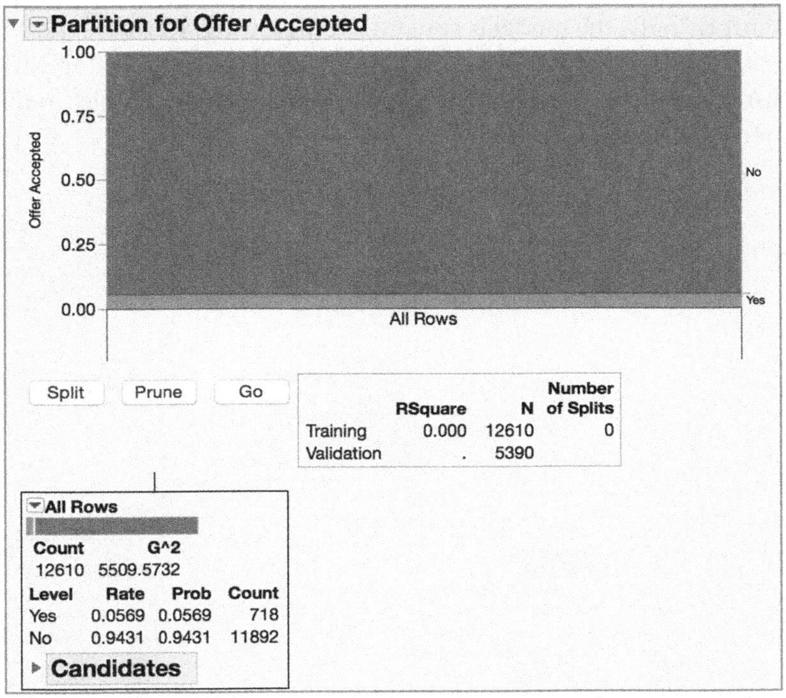

Figure 6.24: Credit, Partition Initial Candidate Splits

Candidates

Term	Candidate G^2		LogWorth	Cut Point
Reward	56.1777539		13.03134984	Cash Back,Points
Mailer Type	120.0291291		27.20524329	Letter
Income Level	55.0386480		12.76925799	Low
# Bank Accounts Open	1.2146267		0.37255410	3
Overdraft Protection	0.0299123		0.06414555	No
Credit Rating	268.2835802	*	61.76015120	Low
# Credit Cards Held	0.1365512		0.02136185	4
# Homes Owned	0.9362771		0.29585750	3
Household Size	3.3771571		0.49301450	2
Own Your Home	0.8373776		0.44351917	Yes
Average Balance	4.5268423		1.809401e-95	1736.3
Q1 Balance	5.4320279		6.233225e-44	529
Q2 Balance	7.4859420		1.076731e-30	104
Q3 Balance	4.8152547		4.205862e-51	11
Q4 Balance	3.8232924		1.113518e-54	13

The tree after three splits (click **Split** three times) is shown in Figure 6.25.

Not surprisingly, the model is split on **Credit Rating**, **Reward**, and **Mailer Type**. The lowest probability of accepting the offer (0.0196) is **Credit Rating(Medium, High)** and **Reward(Cash Back, Points)**. The highest probability (0.1473) is **Credit Rating(Low)** and **Mailer Type(Postcard)**.

Figure 6.25:   Credit, Partition after Three Splits

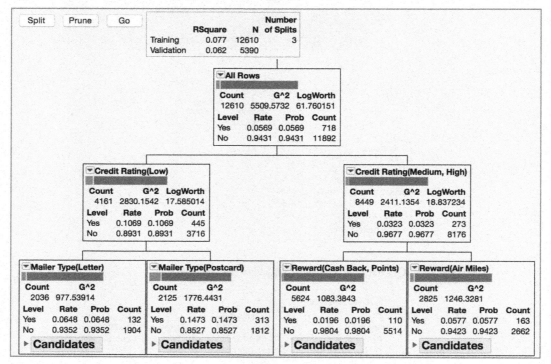

After each split, the model *RSquare* (or, *Entropy RSquare*) updates (this is shown at the top of Figure 6.25). Remember that RSquare is a measure of how much variability in the response is being explained by the model. Without a validation set, we can continue to split until the minimum split size is achieved in each node. (The minimum split size is an option under the top red triangle, which is set to 5 by default.) However, additional splits are not necessarily beneficial and lead to more complex and potentially over-fit models.

Since we have a validation portion, we click **Go** to automate the tree-building process. The tree with the maximum validation RSquare has 15 splits. (See Figure 6.26. Note that the *y*-axis has been re-scaled.) The line for RSquare for the validation data is at its maximum value at 15 splits.

Figure 6.26: Credit, Split History after Fifteen Splits

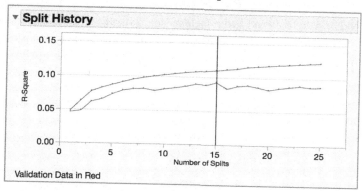

To summarize which variables are involved in the 15 splits, we turn on **Column Contributions** (from the top red triangle). This table indicates which variables are most important in terms of the overall contribution to the model (see Figure 6.27).

Credit Rating, **Mailer Type**, **Reward**, and **Income Level** contribute the most to the model. Several of the variables, including the five balance variables, are not involved in any of the splits.

Figure 6.27: Credit, Split History after Fifteen Splits

▼ **Column Contributions**

Term	Number of Splits	G^2		Portion
Credit Rating	4	319.036751		0.5255
Mailer Type	3	121.564427		0.2002
Reward	4	106.213943		0.1750
Income Level	2	39.8036153		0.0656
# Credit Cards Held	1	13.40561		0.0221
Overdraft Protection	1	7.08462348		0.0117
# Bank Accounts Open	0	0		0.0000
# Homes Owned	0	0		0.0000
Household Size	0	0		0.0000
Own Your Home	0	0		0.0000
Average Balance	0	0		0.0000
Q1 Balance	0	0		0.0000
Q2 Balance	0	0		0.0000
Q3 Balance	0	0		0.0000
Q4 Balance	0	0		0.0000

The Confusion Matrix

One overall measure of model accuracy is the *Misclassification Rate* (select **Fit Details** from the top red triangle). The misclassification rate for our validation data is 0.0573, or 5.73%. The numbers behind the misclassification rate can be seen in the *confusion matrix* (Figure 6.28). We focus on the misclassification rate and confusion matrix for the validation data. Since these data were not used in building the model, this provides a better indication of how well the model classifies **Offer Accepted**.

Figure 6.28:   Credit, Fit Details with Confusion Matrix

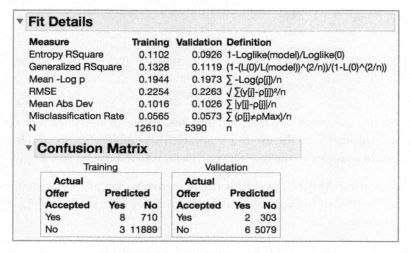

There are four possible outcomes in our classification:

- An accepted offer is correctly classified as an accepted offer.
- An accepted offer is misclassified as not accepted.
- An offer that was not accepted is correctly classified as not accepted.
- An offer that was not accepted is misclassified as accepted.

One observation is that there were few cases where the model predicted that the offer would be accepted. (See the "2" value in the **Yes** column in the validation confusion matrix in Figure 6.28.) When the target variable is unbalanced (i.e., there are far more observations in one level than the other), the model that is fit will usually result in probabilities that are small for the under-represented category.

In this case, the overall rate of **Yes** (offer accepted) is 5.68%, which is close to the misclassification rate for this model. However, when we examine the **Leaf Report** for the fitted model (Figure 6.29), we see that there are terminal branches in the tree that have much richer concentrations of **Offer Accepted** = **Yes** than the overall average rate. Note that the results in the **Leaf Report** are for the training data.

Figure 6.29: Credit, Leaf Report for Fitted Model

Leaf Report

Response Prob

Leaf Label	Yes	No
Credit Rating(Low)&Mailer Type(Letter)	0.0648	0.9352
Credit Rating(Low)&Mailer Type(Postcard)&Income Level(Low)&# Credit Cards Held<4	0.2056	0.7944
Credit Rating(Low)&Mailer Type(Postcard)&Income Level(Low)&# Credit Cards Held>=4	0.6738	0.3262
Credit Rating(Low)&Mailer Type(Postcard)&Income Level(Medium, High)	0.1244	0.8756
^^&Credit Rating(Medium)&Mailer Type(Letter)&Reward(Cash Back)	0.0044	0.9956
^^&Credit Rating(Medium)&Mailer Type(Letter)&Reward(Points)	0.0246	0.9754
^&Reward(Cash Back, Points)&Credit Rating(Medium)&Mailer Type(Postcard)&Income Level(Low)	0.0858	0.9142
^^&Credit Rating(Medium)&Mailer Type(Postcard)&Income Level(Medium, High)&Reward(Cash Back)	0.0187	0.9813
^^&Credit Rating(Medium)&Mailer Type(Postcard)&Income Level(Medium, High)&Reward(Points)	0.0455	0.9545
^^&Credit Rating(High)&Reward(Cash Back)	0.0049	0.9951
^^&Credit Rating(High)&Reward(Points)	0.0146	0.9854
^&Reward(Air Miles)&Mailer Type(Letter)&Credit Rating(Medium)	0.0529	0.9471
^&Reward(Air Miles)&Mailer Type(Letter)&Credit Rating(High)	0.0213	0.9787
^&Reward(Air Miles)&Mailer Type(Postcard)&Credit Rating(Medium)	0.1012	0.8988
^&Reward(Air Miles)&Mailer Type(Postcard)&Credit Rating(High)&Overdraft Protection(No)	0.0450	0.9550
^&Reward(Air Miles)&Mailer Type(Postcard)&Credit Rating(High)&Overdraft Protection(Yes)	0.1126	0.8874

Response Counts

Leaf Label	Yes	No
Credit Rating(Low)&Mailer Type(Letter)	132	1904
Credit Rating(Low)&Mailer Type(Postcard)&Income Level(Low)&# Credit Cards Held<4	106	409
Credit Rating(Low)&Mailer Type(Postcard)&Income Level(Low)&# Credit Cards Held>=4	8	3
Credit Rating(Low)&Mailer Type(Postcard)&Income Level(Medium, High)	199	1400
^^&Credit Rating(Medium)&Mailer Type(Letter)&Reward(Cash Back)	3	693
^^&Credit Rating(Medium)&Mailer Type(Letter)&Reward(Points)	18	715
^&Reward(Cash Back, Points)&Credit Rating(Medium)&Mailer Type(Postcard)&Income Level(Low)	28	298
^^&Credit Rating(Medium)&Mailer Type(Postcard)&Income Level(Medium, High)&Reward(Cash Back)	10	527
^^&Credit Rating(Medium)&Mailer Type(Postcard)&Income Level(Medium, High)&Reward(Points)	24	503
^^&Credit Rating(High)&Reward(Cash Back)	7	1423
^^&Credit Rating(High)&Reward(Points)	20	1355
^&Reward(Air Miles)&Mailer Type(Letter)&Credit Rating(Medium)	36	645
^&Reward(Air Miles)&Mailer Type(Letter)&Credit Rating(High)	15	690
^&Reward(Air Miles)&Mailer Type(Postcard)&Credit Rating(Medium)	71	630
^&Reward(Air Miles)&Mailer Type(Postcard)&Credit Rating(High)&Overdraft Protection(No)	28	595
^&Reward(Air Miles)&Mailer Type(Postcard)&Credit Rating(High)&Overdraft Protection(Yes)	13	102

The fitted model has probabilities of **Offer Accepted** = **Yes** in the range [0.0044, 0.6738]. Recall that when JMP classifies rows with the model, it uses a default of Prob > 0.5 to make the decision. In this case, only one of the predicted probabilities of **Yes** is > 0.5, and this one node has only 11 observations: 8 yes and 3 no under **Response Counts** in the bottom table in Figure 6.29. The next highest predicted probability of **Offer Accepted** = **Yes** is 0.2056. As a result, all other rows are classified as **Offer Accepted** = **No**.

The ROC Curve

Two additional measures of accuracy used when building classification models are *Sensitivity* and *(1-Specificity)*. Sensitivity is the true positive rate. In our example, this is the ability of our model to correctly classify **Offer Accepted** as **Yes**. The second measure, (1-Specificity), is the false positive rate. In this case, a false positive occurs when an offer was not accepted, but was classified as **Yes** (accepted).

Instead of using the default decision rule of *Prob > 0.5*, we examine the decision rule *Prob > T*, where we let the decision threshold T range from 0 to 1. We plot the Sensitivity (on the *y*-axis) versus the (1-Specificity) (on the *x*-axis) for each possible threshold value. This creates a *Receiver Operating Characteristic (ROC)* curve. The ROC curve can be displayed by selecting that option from the top red triangle in the Partition report.

Figure 6.30:   Credit, ROC Curve for Offer Accepted

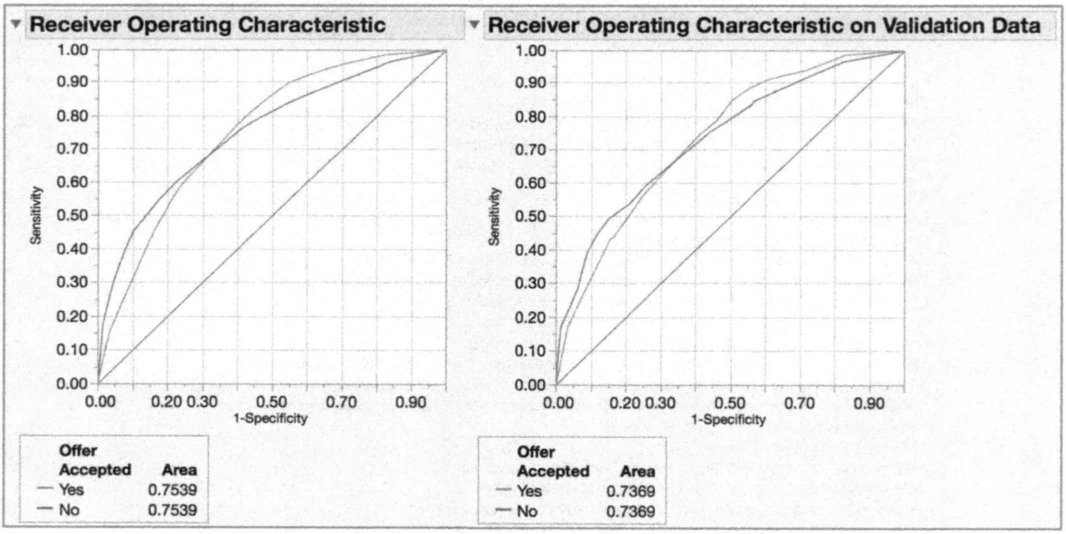

Conceptually, what the ROC curve is measuring is the ability of the predicted probability formulas to rank an observation. Here, we just focus on the **Yes** outcome for the **Offer Accepted** response variable. We save the probability formula to the data table, and then sort the data table from highest to lowest probability. If this probability model can correctly classify the outcomes for **Offer Accepted**, we would expect to see more **Yes** response values at the top (where the probability for **Yes** is highest) than **No** responses.

Similarly, at the bottom of the sorted table, we would expect to see more **No** than **Yes** response values.

Constructing an ROC Curve

Here is a practical algorithm to quickly draw an ROC curve after the table has been sorted by the predicted probability. We walk through the algorithm for **Offer Accepted = Yes**, but this is done automatically in JMP for each response category.

For each observation in the sorted table, starting at the observation with the highest probability **Offer Accepted = Yes**:

- If the observed response value is **Yes**, then a vertical line segment (increasing, along the Sensitivity axis) is drawn. The length of the line segment is 1/(total number of **Yes** responses in the table).

- If the observed response value is **No**, then a horizontal line segment (increasing, along the 1-Specificity axis) is drawn. The length of the line segment is 1/(total number of "No" responses in the data table).

Simple ROC Curve Examples

We use a simple example to illustrate. Suppose we have a data table with only 8 observations. We sort these observations from high to low based on the probability that the **Outcome = Yes**. The sorted actual response values are **Yes, Yes, Yes, No, Yes, Yes, No**, and **Yes**. This results in the ROC curve on the left of Figure 6.31 (arrows are added to show the steps in the ROC curve construction). The first 3 line segments are drawn up because the first three sorted values have **Outcome = Yes.**

Now, suppose that we have a different probability model that we use to rank the observations, resulting in the sorted outcomes **Yes, No, Yes, No, No, Yes, No**, and **Yes**. The ROC curve for this situation is shown on the right of Figure 6.31. The first ROC curve moves "up" faster than the second curve. This is an indication that the first model is doing a better job of separating the **Yes** responses from the **No** responses, based on the predicted probability.

Figure 6.31:   ROC Curve Examples

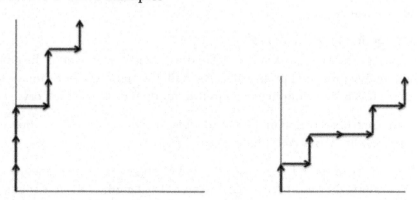

Referring back to the example ROC curve in Figure 6.30, we see that JMP Pro has also displayed a diagonal reference line on the chart, which represents the *Sensitivity = 1-Specificity* line. If a probability model cannot sort the data into the correct response category, then it may be no better than just sorting at random. In this case, the ROC curve for a "random ranking" model would be similar to this diagonal line. A model that sorts the data perfectly, with all the **Yes** responses at the top of the sorted table, would have an ROC Curve that goes from the origin of the graph straight up to sensitivity = 1, then straight over to 1-specificity = 1. A model that sorts perfectly can be made into a classifier rule that classifies perfectly. That is, a classifier rule that has a sensitivity of 1.0 and 1-specificity of 0.0.

The *area under the curve*, or *AUC* (labeled **Area** in Figure 6.30) is a measure of how well our model sorts the data. The diagonal line, which would represent a random sorting model, has an AUC of 0.5. A perfect sorting model has an AUC of 1.0. The area under the curve for **Offer Accepted = Yes** is 0.7369 (see Figure 6.30), indicating that the model predicts better than the random sorting model.

The Lift Curve

Another measure of how well a model can sort outcomes is the model *lift*. As with the ROC curve, we examine the table that is sorted in descending order of the predicted probability. For each sorted row, we calculate the sensitivity, and we divide that by the proportion of values in the table where **Offer Accepted = Yes**. This value is the model lift.

Lift is a measure of how much "richness" in the response we achieve by applying a classification rule to the data. A **Lift Curve** plots the **Lift** (on the *y*-axis) against the

Portion (on the *x*-axis). Again, consider the data table that has been sorted by the predicted probability of a given outcome. As we go down the table from the top to the bottom, the portion is just the relative position of the row that we are considering. The top 10% of the rows in the sorted table corresponds to a portion of 0.1, the top 20% of the rows corresponds to a portion of 0.2, and so on. The lift for **Offer Accepted = Yes**, for a given portion, is simply the proportion of **Yes** responses in this portion, divided by overall proportion of **Yes** responses in the entire data table.

The higher the lift at a given portion, the better our model is at correctly classifying the outcome within this portion. For **Offer Accepted = Yes**, the lift at Portion = 0.15 is roughly 2.5 (see Figure 6.32). This means that in rows in the data table that correspond to the top 15% of the model's predicted probabilities, the number of actual **Yes** outcomes is 2.5 times higher than we would expect if we had just chosen 15% of the rows from the data set at random. If the model is not sorting the data well, then the lift will hover at around 1.0 across all of the portion values.

Figure 6.32:   Lift Curve for Offer Accepted

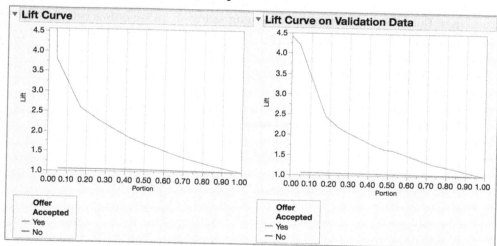

Lift provides another measure of how good our model is at classifying outcomes, and it is particularly useful if the overall predicted probabilities are lower than 0.5 for the outcome that we wish to predict. Even though, in this example, the majority of the predicted probabilities of **Offer Accepted = Yes** were less than 0.2, the lift curve indicates that there are threshold values that we could use with the predicted probability model to create a classifier rule that will be better than just guessing at random. This rule can be used to identify portions of our data that have a much richer number of customers that are likely to accept an offer.

For categorical response models, the misclassification rate, the confusion matrix, the ROC curve, and the lift curve all provide measures of model accuracy, and each of these should be used to assess the quality of the prediction model.

Case Summary

This model was created for explanatory rather than predictive purposes. We wanted to understand the characteristics of customers most likely to accept a credit card offer. Based on this model, the four most important factors are: **Credit Rating**, **Mailer Type**, **Reward**, and **Income Level**. How can this information be used? We'll explore this question in an exercise at the end of this chapter.

Example 2:   Printing Press Yield

A company uses a rotogravure printing process to print high volume commercial magazines. Rotogravure printing involves engraving an image onto a cylinder (see Figure 6.33, image used with permission of the Gravure Association of the Americas, www.gaa.org). The cylinder is run through an ink bath, excess ink is removed, and the image is transferred to the paper. The images are removed from the cylinders after each job and the cylinders are reused.

Figure 6.33:   The Rotogravure Process

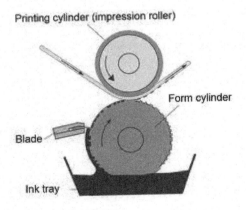

Printing cylinder (impression roller)

Form cylinder

Blade

Ink tray

To stay competitive in the online era, the company must improve its profit margins. For years, the company has been struggling with inefficiencies in the printing process and poor yields. A major contributor is a defect called banding, which consists of grooves that appear in the cylinder during a print run. Banding can result in ruined product, rework, long periods of downtime, unhappy customers, and can be costly to the company.

The Study

Data were compiled on 539 previous production runs, with the goal of identifying the conditions that may lead to banding. The data, a description of the variables, and reference information can be found in the UC Irvine Machine Learning Repository (available at http://archive.ics.uci.edu/ml/datasets/Cylinder+Bands).

In this example, some of the original variables that aren't useful for identifying potential causes of banding and low yields (such as customer number and job number) have been omitted. An additional variable measuring batch yield has been created for illustration.

The Data   PrintingPressYieldBBM.jmp

The data table consists of information on the 539 runs from the rotogravure printing process. There are two potential response variables, **Banding?** and **BatchYield**, and 31 variables representing press characteristics and measurements and customer information.

> **Banding?**: Occurrence of banding (Yes or No) during the production run

> **BatchYield:** The percent of good (usable) product in the job

A summary of the potential predictors, produced using **Cols > Columns viewer**, is shown in Figure 6.34.

Applying the Business Analytics Process

Define the Problem

We want to build a model that will help identify potential causes of poor yield. Since banding is just one of the many causes of yield loss, we focus on the continuous response **BatchYield**. We have a large number of potential predictor variables, many of which are categorical. So we use the **Partition** platform to build a regression tree.

Banding is known to be a serious issue. In an exercise at the end of this chapter, we will ask you to build a classification tree for **Banding?**.

Prepare for Modeling

We follow the usual approach of exploring our data one variable at a time, two variables at a time and then many variables at a time prior to building a regression tree. In this section, we only briefly explore the data to provide background information and context for the case study. As always, we encourage you to more thoroughly explore this data on your own.

Since we have a large data set in terms of the number of variables, we start by creating numerical summaries of each variable using **Cols > Columns Viewer**. (Select **Data Table View** to create a data table with the summary information.)

The first 12 variables (in Figure 6.34) are categorical, mostly with two or three levels. Nearly all of the variables are missing values. A peek at the minimum and maximum values for the continuous variables indicates another potential problem with data quality. For some variables (for example, **Press Speed**), the minimum value is 0. Is this a feasible value, or should the value be coded as missing? The **Partition** platform provides the **Informative Missing** option to handle missing values (see the discussion earlier in the chapter). We ask you to consider potential issues caused by missing values and other data quality concerns in an exercise at the end of the chapter.

Figure 6.34: Printing Press Yield Study – Potential Predictors

	Columns	N	N Missing	N Categories	Min	Max	Mean	Std Dev
1	grain screened	491	48	2	•	•	•	•
2	proof on ctd ink	483	56	2	•	•	•	•
3	blade mfg	480	59	2	•	•	•	•
4	paper type	539	0	3	•	•	•	•
5	ink type	539	0	3	•	•	•	•
6	direct steam	515	24	2	•	•	•	•
7	solvent type	485	54	3	•	•	•	•
8	type on cylinder	521	18	2	•	•	•	•
9	press type	539	0	4	•	•	•	•
10	cylinder size	536	3	3	•	•	•	•
11	paper mill location	384	155	5	•	•	•	•
12	plating tank	521	18	2	•	•	•	•
13	proof cut	485	54	•	25	72.5	45.036082474	9.0440974949
14	viscosity	534	5	•	35	72	50.949438202	8.0588402314
15	caliper	512	27	•	0.133	0.533	0.2758574219	0.0694738486
16	ink temperature	537	2	•	11.2	24.5	15.359795158	1.2779931778
17	humidity	538	1	•	57	105	78.539033457	7.7346207467
18	roughness	509	30	•	0.05625	1.25	0.7243334971	0.1935045789
19	blade pressure	476	63	•	16	70	30.907563025	9.1234726506
20	varnish pct	484	55	•	0	35.8	5.7805785124	6.8454413415
21	press speed	529	10	•	0	2600	1823.3591682	328.70603681
22	ink pct	484	55	•	41	76.9	55.638533058	5.5614101944
23	solvent pct	484	55	•	22	53.4	38.567644628	3.5014309157
24	ESA Voltage	483	56	•	0	16	1.3193581781	2.4626047595
25	ESA Amperage	485	54	•	0	6	0.0381443299	0.4154663065
26	wax	533	6	•	0	3.1	2.4009380863	0.5455467888
27	hardener	532	7	•	0	3	0.987593985	0.3679564381
28	roller durometer	485	54	•	28	60	34.775257732	4.5057362642
29	current density	532	7	•	30	45	39.056390977	2.3531476178
30	anode space ratio	532	7	•	83.33	117.86	103.03279135	5.0089486915
31	chrome content	536	3	•	90	100	99.598880597	1.8538967943

The average yield is only 79.42% (Figure 6.35), and many job yields are extremely low. Roughly 25% were less than 74%, and only 25% were over 90%.

Figure 6.35: Press, BatchYield Distribution

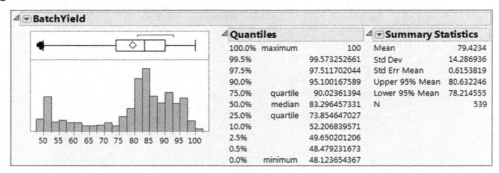

A subset of the variables is shown in Figure 6.36. The jobs with the lower yields are selected. Some of the variables seem to stand out as having more of the low yields. For example, there appears to be more shading when **grain screened**=YES than NO, and when **paper type** is UNCOATED. In addition, all cases of the **paper type**=SUPER are shaded.

Figure 6.36: Press, BatchYield, and Subset of Categorical and Continuous Predictors

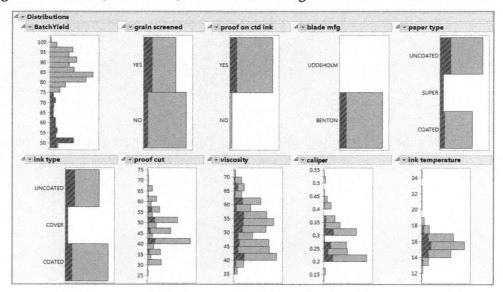

To further explore potential relationships, we use **Fit Y by X** (Figure 6.37). We start with the nominal variables and use **Column Switcher** to efficiently toggle between variables.

We do the same with the continuous predictors, and some interesting patterns start to emerge. (**Fit Line** and **Nonpar Density** have been selected in Figure 6.38.) For example, there appears to be a positive relationship between **press speed** and **BatchYield**, but there are two groupings of points. A large cluster with **BatchYield** values roughly 80-90 percent, and another smaller grouping at roughly 50%.

Figure 6.37: Press, BatchYield versus Grain Screened with Column Switcher

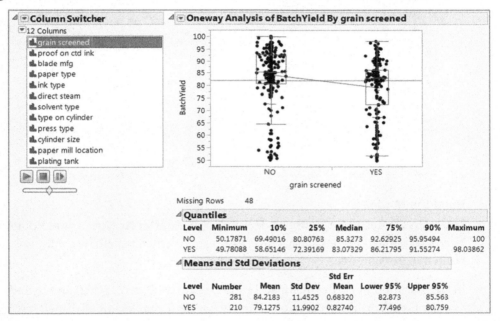

Level	Minimum	10%	25%	Median	75%	90%	Maximum
NO	50.17871	69.49016	80.80763	85.3273	92.62925	95.95494	100
YES	49.78088	58.65146	72.39169	83.07329	86.21795	91.55274	98.03862

Means and Std Deviations

Level	Number	Mean	Std Dev	Std Err Mean	Lower 95%	Upper 95%
NO	281	84.2183	11.4525	0.68320	82.873	85.563
YES	210	79.1275	11.9902	0.82740	77.496	80.759

Figure 6.38:   Press, BatchYield versus Grain Screened with Column Switcher

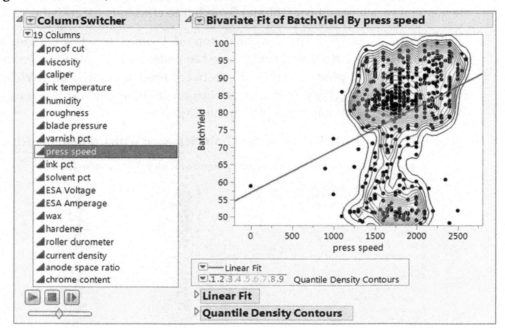

Build the Model

In the **Credit Card Marketing** case, our response, **Offer Accepted**, was nominal, and we built a classification tree. In this case, our response, **BatchYield**, is continuous, so we will build a regression tree.

In classification trees, the likelihood ratio chi-square statistic (G^2) and the adjusted *p*-value (LogWorth) are used to determine optimal split locations. When the response is continuous, the sum of squares (SS) due to the differences between means is used rather than the G^2. But the optimal split location is still determined by maximizing the LogWorth statistic. In the case of a continuous response, the fitted values are the means within the two groups.

We use the **Partition** platform, with **BatchYield** as **Y, Response** and all of the predictors as **X, Factor**, and accept the default settings. The initial output is shown in Figure 6.39.

Figure 6.39: **Press, Initial Partition View**

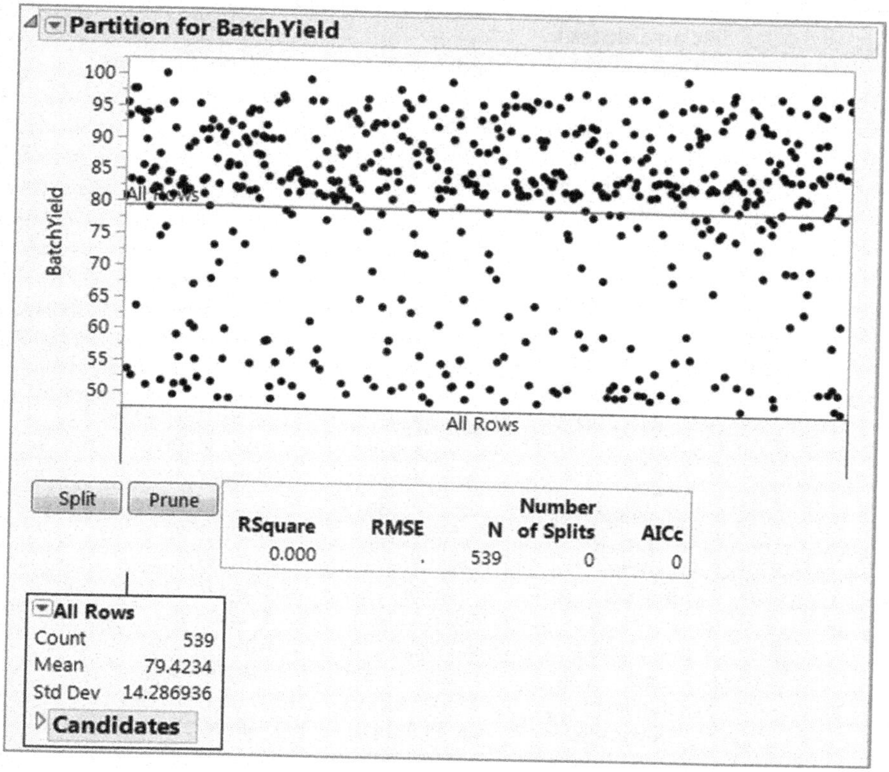

The best candidate for the first split, based on the LogWorth, is **ink pct** (Figure 6.40).

Figure 6.40:   Press, Initial Candidates

⊿ Candidates			
Term	**Candidate SS**	**LogWorth**	**Cut Point**
grain screened	37775.91523	65.7361635	
proof on ctd ink	46184.66331	90.9039658	
blade mfg	45089.39172	87.2556263	
paper type	20403.96249	28.7173303	SUPER
ink type	7994.72711	9.9796273	COVER
direct steam	21179.23971	30.0607744	
solvent type	49119.77139	106.3900767	
type on cylinder	6505.72217	8.0290169	,NO
press type	11623.80902	15.0628782	Albert70,WoodHoe70
cylinder size	3713.62576	4.1923348	,SPIEGEL,CATALOG
paper mill location	28286.53880	46.9999376	,SouthUS
plating tank	208.14097	0.2795366	1911,
proof cut	36041.79761	74.2747934	25
viscosity	3476.46649	3.5834675	37
caliper	3032.02355	3.1336981	0.167
ink temperature	1310.03550	0.9012423	18.3
humidity	2739.26276	2.6037272	76
roughness	4951.25079	5.8073495	0.75
blade pressure	5105.38083	5.8759584	22
varnish pct	50103.70362	128.5718616	0
press speed	10779.02622	14.8004649	2100
ink pct	50111.96175 *	128.6135761	68.2
solvent pct	50103.70362	128.5718729	22
ESA Voltage	46109.88191	110.7980398	0
ESA Amperage	49119.77139	115.1153559	0
wax	4080.26826	4.4383506	0.7
hardener	5839.54058	6.9963575	2
roller durometer	49119.77139	121.7755758	28
current density	2204.79053	2.3349512	37
anode space ratio	1087.90408	0.6745484	114.28
chrome content	4771.36365	6.2076144	100

The tree after one split is shown in Figure 6.41 (again, we use the **Split** button). When **ink pct** is < 68.2, the average yield is 82.77. In contrast, when **ink pct** is >=68.2, or the value of **ink pct** is missing, the average yield is only 51.66. In this branch, the standard deviation of yield is low, and the points in the partition graph have little vertical spread, indicating that yields at the higher ink percentages (or missing) are consistently poor.

A peek at the two sets of candidates indicates that the highest LogWorth is in the **ink pct is <68.2 or not missing** branch, and the split will be on **press type**. The resulting tree is shown in Figure 6.42.

Figure 6.41: Press after First Split

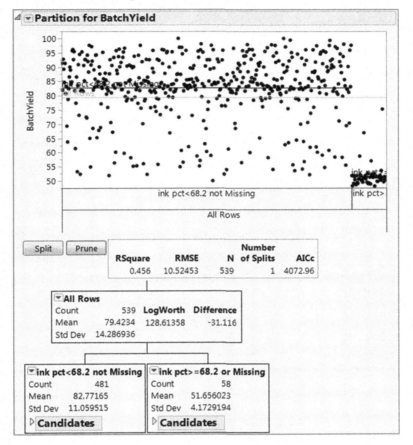

We continue to split. The resulting tree after five splits is shown in Figure 6.43. The next four splits were below **ink pct <68.2 or not missing**. Within this branch, we see the lowest mean yields on the far left and the highest on the far right. The key variables are **ink pct**, **press type**, **type on cylinder**, and **paper mill location**.

Figure 6.42: Press after Two Splits

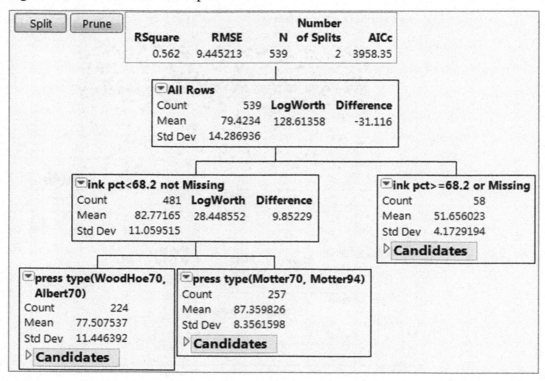

Figure 6.43:   Press after Five Splits

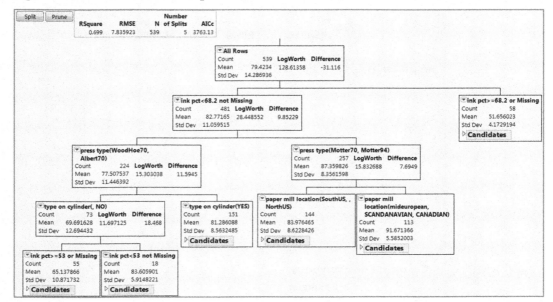

The RSquare for the model has increased from 0.456 after one split to 0.699 (top, in Figure 6.43). After only five splits, we've started to get a sense for the factors and conditions resulting in poor yields. We can continue to split, or we can explore branches of the tree interactively (click on a red triangle for a node and select **Split Here** or **Split Specific**).

In this example, we will continue to split to see what else we can learn as the tree grows. But, with many splits, the tree can become difficult to navigate. Before splitting further, we turn off the tree view. (Under the top red triangle, select **Display Options > Show Tree**. Note that a condensed version of the tree can be found under **Small Tree View**.) We also turn on **Split History** and **Column Contributions**, which provide summaries of how the RSquare changes as we build the tree and the variables involved in the splits.

After 10 splits, the RSquare is 0.765, and we see 6 key variables emerge under **Column Contributions** (Figure 6.44).

Figure 6.44:   Press after 10 Splits

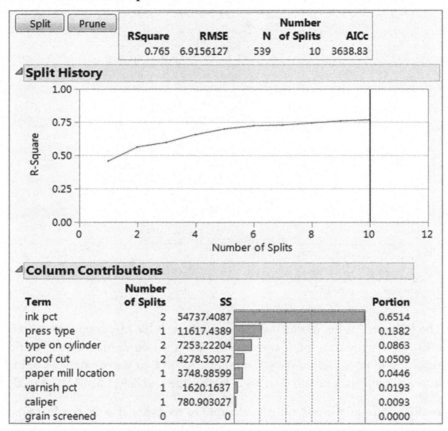

The leaf report provides a summary of the tree. In Figure 6.45, we've sorted the leaf report on the **Mean** (right-click on the report and select **Sort by Column**). It's easy to see that high or missing **ink pct** is a problem, with an average yield of only 51.66. But other conditions also result in low yields. For example, in the fourth row, which has 55 observations, we see that **ink pct**, **press type**, **type on cylinder**, **varnish pct**, and **caliper** are involved, and the average yield is only 65.14.

Figure 6.45:   Press after 10 Splits

Leaf Label	Mean	Count
ink pct>=68.2 or Missing	51.6560232	58
ink pct<68.2 not Missing&press type(WoodHoe70, Albert70)&type on cylinder(YES)&varnish pct<8.3&proof cut<27.5 or Missing	60.879927	5
ink pct<68.2 not Missing&press type(Motter70, Motter94)&paper mill location(SouthUS, , NorthUS)&proof cut<30 or Missing	61.8415925	5
ink pct<68.2 not Missing&press type(WoodHoe70, Albert70)&type on cylinder(, NO)&ink pct>=53 or Missing	65.1378663	55
ink pct<68.2 not Missing&press type(WoodHoe70, Albert70)&type on cylinder(YES)&varnish pct>=8.3&caliper>=0.5 or Missing	74.3448748	5
ink pct<68.2 not Missing&press type(WoodHoe70, Albert70)&type on cylinder(YES)&varnish pct<8.3&proof cut>=27.5 not Missing	80.0034589	99
ink pct<68.2 not Missing&press type(WoodHoe70, Albert70)&type on cylinder(, NO)&ink pct<53 not Missing	83.6059011	18
ink pct<68.2 not Missing&press type(Motter70, Motter94)&paper mill location(SouthUS, , NorthUS)&proof cut>=30 not Missing	84.7726837	139
ink pct<68.2 not Missing&press type(Motter70, Motter94)&paper mill location(mideuropean, SCANDANAVIAN, CANADIAN)&type on cylinder(NO)	87.2137893	25
ink pct<68.2 not Missing&press type(WoodHoe70, Albert70)&type on cylinder(YES)&varnish pct>=8.3&caliper<0.5 not Missing	87.565069	42
ink pct<68.2 not Missing&press type(Motter70, Motter94)&paper mill location(mideuropean, SCANDANAVIAN, CANADIAN)&type on cylinder(, YES)	92.9377226	88

We continue splitting. The R-Square in the **Split History** seems to level off somewhat after 9 or 10 splits (Figure 6.46).

Figure 6.46:   Press, Split History after 15 Splits

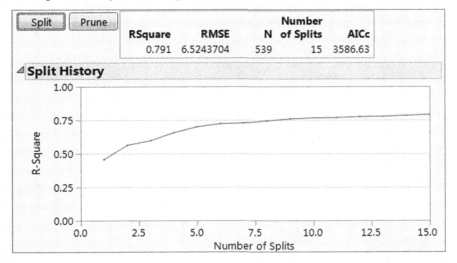

How do we know when to stop splitting? As we have seen, using validation data helps identify the optimal number of splits to balance over-fitting (too many splits) and under-fitting (not enough splits).

In this example, our objectives are purely exploratory in nature. We're interested in identifying important factors and problematic conditions rather than building a model to predict yield. So one approach is to keep splitting and see what "falls out." A second approach is to limit the number of splits by changing the minimum split size (under the top red triangle). This will prevent splitting unless there is a minimum number of observations in a node.

We proceed with the first approach (we explore the minimum split size in an exercise). After 40 splits, a few other variables enter the picture, but the splits continue to be dominated by the same six variables (Figure 6.47):

- **ink pct**
- **press type**
- **type on cylinder**
- **proof cut**
- **paper mill location**
- **varnish pct**

Figure 6.47:   Press, Column Contributions after 40 Splits

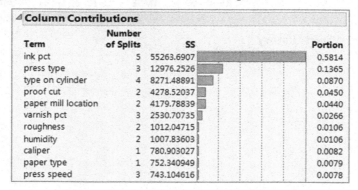

Term	Number of Splits	SS		Portion
ink pct	5	55263.6907		0.5814
press type	3	12976.2526		0.1365
type on cylinder	4	8271.48891		0.0870
proof cut	2	4278.52037		0.0450
paper mill location	2	4179.78839		0.0440
varnish pct	3	2530.70735		0.0266
roughness	2	1012.04715		0.0106
humidity	2	1007.83603		0.0106
caliper	1	780.903027		0.0082
paper type	1	752.340949		0.0079
press speed	3	743.104616		0.0078

Case Summary

What do we do with this knowledge? While we have evidence that these variables are involved in low yields, and that yields are worse than others under some conditions, we don't have *causal* evidence. It could be that some other unmeasured variable or condition is the root cause of the problem. In industrial situations such as this, it may be possible to design a formal experiment to directly study these variables. In a designed experiment, the settings of key factors are systematically changed, and other variables are held constant, while response variables are measured. In this situation, a designed experiment may allow us to develop a better understanding of these six potential root causes of poor yield, and to identify optimal settings of the factors to maximize yield (see Gaudard et al., 2006).

Summary

In this chapter, we have introduced decision trees primarily from the perspective of exploratory modeling, where the goals have been to identify important variables. But, decision trees can also be used to develop accurate predictive models. In Chapter 8, we'll discuss methods for using cross-validation with partition trees to help build the "best-sized" model. In Chapter 9, we'll introduce Bootstrap Forests and Boosted Trees, two additional tree-based methods.

Exercises

Exercise 6.1: Use the **Lost Sales BBM.jmp** data set for this exercise. In Chapter 5, we created a logistic regression model for **Status**. At the beginning of this chapter, we used this data to introduce classification trees and build a 5-split model.

Build a classification tree for **Status**, and see if you can improve upon the model shown.

  a. How many splits are in your final tree?
  b. What is your final model?
  c. What is the misclassification rate for your model?
  d. What is the probability that **Status** = **Won** for an after market part with a quoted time to delivery of 37 days and a quoted price of $1000?

Exercise 6.2: In the Credit Card Marketing example in this chapter, we identified four important characteristics of customers likely to accept a credit card offer.

  a. Why was this an exploratory model rather than a predictive model?
  b. How can this information be used by the company? What is the potential value of identifying these characteristics?

Exercise 6.3: Use the **Titanic Passengers BBM.jmp** data set for this exercise. Recall that in Chapter 5 we created a logistic model to predict survival.

Build a classification tree for **Survived** using the available variables. (Refer to Chapter 5 for information on the data and definitions of each variable.) Use **Column Contributions** and **Split History** to determine the optimal number of splits. Save your results; we'll use your final model in Exercise 6.4. Do not use a validation set for this exercise.

    a.   How many splits are in your final tree?

    b.   Which variables are the largest contributors?

    c.   What is your final model?

    d.   What is the misclassification rate for this model? Is the model better at predicting survival or nonsurvival?

    e.   What is the area under the ROC curve for **Survived**? Interpret this value. Does the model do a better job of classifying survival than a random model?

    f.   What is the lift for the model at portion = 0.1 and at portion = 0.25? Interpret these values.

Exercise 6.4: Refer to the final logistic model in Chapter 5 for the **Titanic Passengers BBM.jmp**.

    a.   What was the misclassification rate for the logistic model?

    b.   Compare the misclassification rates for the logistic model and the partition model created in Exercise 6.3. Which model is better? Why?

    c.   Which model would be easier to explain to a non-technical person? Why?

Exercise 6.5: Use the **PrintingPressYieldBBM.jmp** data set for this exercise. In this chapter, we fit a regression tree for **BatchYield**. We ignored potential data quality issues, and let JMP handle missing values for us.

    a.   Re-create the tree in Figure 6.43, and save the prediction formula to the data table. In the data table, open the formula for this new column. Interpret the formula. How is the model handling the missing values?

    b.   Click on the red triangle next to **ink pct < 68.2 or Not Missing** in the regression tree, and select **Split Specific**. Split on **type on cylinder**. How is JMP handling the missing values for **type on cylinder**, which is a nominal variable? Again, save the prediction formula to the data table and interpret the formula.

    c.   As we saw in Figure 6.34, some of the continuous variables have minimum values of zero. Is zero a valid minimum value for these variables? Why or why not? If these values are not valid, what impact do they have on the modeling process?

    d.   Explore the distributions of the continuous variables. Are there any other unusual observations?

e. What can you, as a modeler, do to "fix" or address data quality issues? List several actions that you might take for this example.

Exercise 6.6: Use the **PrintingPressYieldBBM.jmp** data set for this exercise. Repeat the analysis shown in this chapter, but instead fit a classification tree for **Banding?**. Do not use validation for this exercise.

a. First, change the value ordering for **Banding?** so that you model the probability that **Banding?=Yes**.

b. After 10 splits, which variables are the biggest contributors?

    i. What is the misclassification rate?

    ii. Which category is misclassified most often?

    iii. What is the AUC (area under the ROC curve)?

    iv. What is the Lift at portion = 0.1?

c. Split 5 more times. What is the misclassification rate, the AUC, and the Lift at portion = 0.1?

d. Prune the model all the way back, and change the **Minimum Split Size** to 25. Then split until no further splits are possible. How many times did the tree split? How did this change the column contributions, the AUC and the Lift at portion = 0.1?

References

Bache, K. and M. Lichman. 2013. UCI Machine Learning Repository (http://archive.ics.uci.edu/ml). Irvine, CA: University of California, School of Information and Computer Science. The Printing Yield Data is based on the data available at https://archive.ics.uci.edu/ml/datasets/Cylinder+Bands .

Gaudard, M., P. Ramsey, and M. Stephens. 2006. Interactive Data Mining and Design of Experiments: the JMP Partition and Custom Design Platforms. Available at http://www.jmp.com/en_us/whitepapers/jmp/interactive-data-mining-doe.html.

Sall, J. 2002. "Monte Carlo Calibration of Distributions of Partition Statistics." Available at http://www.jmp.com/en_us/whitepapers/jmp/monte-carlo-calibration.html.

7

Neural Networks

In the News

The technology called *deep learning* has already been put to use in services like Apple's Siri virtual personal assistant, which is based on Nuance Communications' speech recognition service, and in Google's Street View, which uses machine vision to identify specific addresses.

But what is relatively new is the growing speed and accuracy of deep-learning programs, often called artificial neural networks or just "neural nets" for their resemblance to the neural connections in the brain.

"There have been a number of stunning new results with deep-learning methods," said Yann LeCun, a computer scientist at New York University who did pioneering research in handwriting recognition at Bell Laboratories. "The kind of jump we are seeing in the accuracy of these systems is very rare indeed."

Artificial intelligence researchers are acutely aware of the dangers of being overly optimistic. Their field has long been plagued by outbursts of misplaced enthusiasm followed by equally striking declines.

In the 1960s, some computer scientists believed that a workable artificial intelligence system was just 10 years away. In the 1980s, a wave of commercial start-ups collapsed, leading to what some people called the "A.I. winter."

But recent achievements have impressed a wide spectrum of computer experts. In October, for example, a team of graduate students studying with the University of Toronto computer scientist Geoffrey E. Hinton won the top prize in a contest sponsored by Merck to design software to help find molecules that might lead to new drugs. From a data set describing the chemical structure of thousands of different molecules, they used deep-learning software to determine which molecule was most likely to be an effective drug agent. The achievement was particularly impressive because the team decided to enter the contest at the last minute and designed its software with no specific knowledge about how the molecules bind to their targets. The students were also working with a relatively small set of data; neural nets typically perform well only with very large ones.

"This is a really breathtaking result because it is the first time that deep learning won, and more significantly it won on a data set that it wouldn't have been expected to win at," said Anthony Goldbloom, chief executive and founder of

Kaggle, a company that organizes data science competitions, including the Merck contest.

Advances in pattern recognition hold implications not just for drug development but for an array of applications, including marketing and law enforcement. With greater accuracy, for example, marketers can comb large databases of consumer behavior to get more precise information on buying habits. And improvements in facial recognition are likely to make surveillance technology cheaper and more commonplace. (Markoff, John. 2012. "Scientists See Advances in Deep Learning, a Part of Artificial Intelligence," NY Times.com. Available at http://nyti.ms/1xgSEYV.)

Representative Business Problems

Neural Network models can be used for both classification and prediction. Therefore, the types of business problems we have seen that can be modeled with Logistic Regression, Multiple Regression, or Classification /Regression Trees can also be modeled with Neural Networks.

For example, neural network models can be used to:

- Identify fraudulent checks submitted to a bank
- Predict system failures on an assembly line
- Predict flight delays
- Determine which content to display on a website based on mouse clicks
- Predict which books a customer is likely to buy based on previous purchases
- Determine whether an employee is a flight risk

Measuring Success

When used for classification with a categorical outcome (a categorical target variable), performance measures such as the misclassification rate, area under the ROC curve (AUC), and the lift curve are used. When used for predictive modeling with a continuous response, the magnitude of the error measured by RMSE (root mean square error) or Mean Abs Dev (mean absolute deviation) are typically used. For these measures, a lower value is better.

Preview of the End Result

The neural network's flexibility enables these models to be very effective in classification and predictive modeling. However, a disadvantage is that the models can be mathematically complex. Unlike decision trees, which can produce useful and interpretable rules for classification or prediction purposes, neural networks produce no such rules. They are based on hidden functions and transformations, and as a result can be difficult to interpret and explain. Fortunately, additional tools are available with JMP that help you understand and interpret the relationships that are uncovered by neural network models.

Looking Inside the Black Box: How the Algorithm Works

We use the data **Boston Housing BBM NN.jmp** for illustration. The response is **mvalue** (median home value), and three of the predictors are **crim** (crime rate), **indus** (percent of land occupied by nonretail business), and **rooms** (average number of rooms per dwelling). In Figure 7.1, we can see that the relationships between the predictors and the response are somewhat complicated. For example, the relationship between **rooms** and **mvalue** is partly curved (for lower values of rooms) and partly linear (at higher values).

Figure 7.1: Plots of mvalue versus Model Factors with Smooth Curves

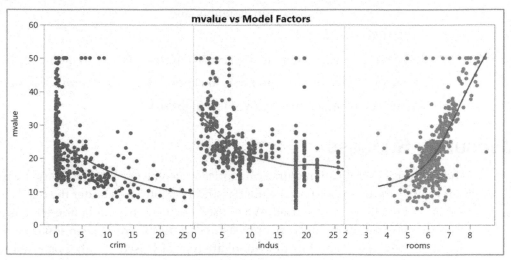

A neural network is a very flexible algorithm that can model complex relationships between inputs and outputs. Each neural network has an input layer, one or more hidden layers, and an output layer. In the neural network in Figure 7.2, the response, **mvalue**, is in the output layer. The input layer consists of the three predictors: **crim**, **indus**, and **rooms.** The input layer is connected to the output layer through a single hidden layer with three nodes.

Figure 7.2: Neural Network Diagram

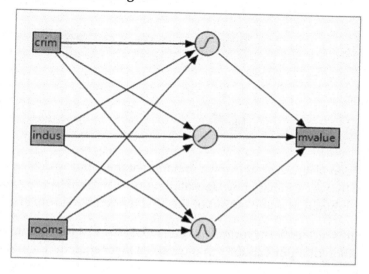

In each node in the hidden layer, a linear combination of the input variables is transformed. The transformation that is applied is denoted with a symbol that indicates the type of transformation.

In JMP Pro, there are three types of transformation functions that can be used in a neural network model: *TanH, Linear,* and *Gaussian*. These are also referred to as *activation functions*. In the standard version of JMP, only the TanH function is available.

- TanH is the hyperbolic tangent function, which is similar in shape to the logistic function that we used for logistic regression models in Chapter 5.

- Linear is similar to constructing a linear regression model. The linear combination of the predictor variables is not transformed.

- Gaussian is a bell-shaped function, which is similar to the normal distribution density function.

Let's look at each hidden node individually in the model from Figure 7.2. The first hidden node is represented by the diagram in Figure 7.3.

Figure 7.3: TanH Node from Neural Network Model

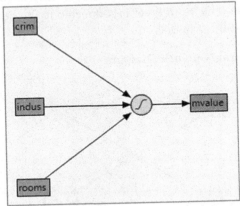

This hidden layer uses a linear combination of the input variables, with weights on each variable that are estimated by **Neural** platform. The TanH() function is applied to this linear combination as the output of this first hidden node, which we label H1_1.

We see this hidden layer formula expressed as a saved formula in Figure 7.4. The linear combination of input variables in the hidden layer contains an intercept and coefficients for each of the predictors. Note that in neural networks, the term *bias* is sometimes used instead of intercept, and the coefficients are sometimes referred to as *weights*.

Figure 7.4: Hidden Layer Formula for First Hidden Node (H1_1)

$$\text{TanH}\left(0.5 * \left(-19.3653496286 + -0.1874591629381 * crim + 0.02985225588797 * indus + 2.87561805010124 * rooms\right)\right)$$

In Figure 7.5, we use a JMP profiler to graphically view this hidden layer formula. Note that the contours for **crim** and **rooms** are similar to the logistic (sigmoidal) function that we saw in Chapter 5.

Figure 7.5:   Profiler View of First Hidden Node (H1_1)

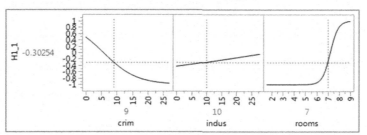

The second node from the hidden layer, H1_2, is represented in the diagram in Figure 7.6. The formula and profiler views of this node's function are shown in Figures 7.7 and 7.8. The output from the hidden layer is simply a linear combination of the inputs.

Figure 7.6:   Linear Node from Neural Network Model

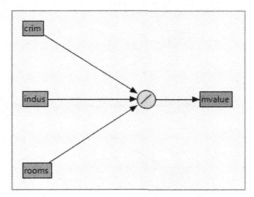

Figure 7.7:   Hidden Layer Formula for Second Hidden Node (H1_2)

-10.223613036057 + -0.198377228052 * *crim* + -0.174135977076 * *indus* + 1.96126111034954 * *rooms*

Note that the contours for each of the factors in the profiler (Figure 7.8) are linear.

Figure 7.8: Profiler View of Second Hidden Node (H1_2)

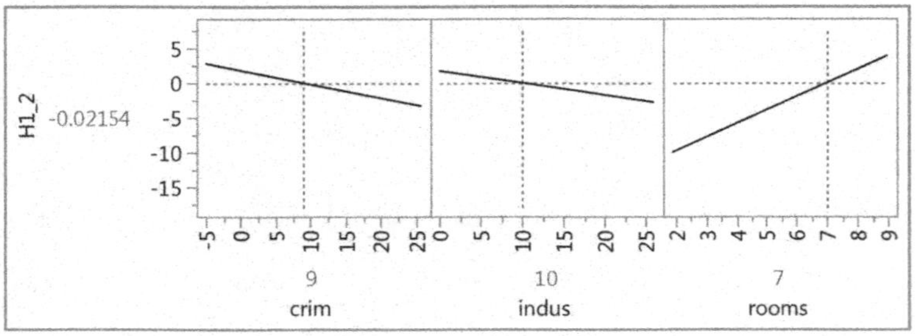

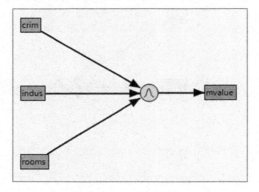

The third node from the hidden layer, H1_3, is represented in the diagram in Figure 7.9, and the formula and profiler view of this node's function is shown in Figures 7.10 and 7.11.

Figure 7.9: Gaussian Node from Neural Network Model

Figure 7.10: Hidden Layer Formula for Third Hidden Node (H1_3)

$$\text{Exp}\left(-\left[0.5 * \left(-21.985086592206 + -0.0597999846762 * crim + 0.07820383855156 * indus + 3.11543189367497 * rooms\right)^2\right]\right)$$

Note that the contours in Figure 7.11 are mounded in shape, similar to the normal distribution density function.

Figure 7.11: Profiler View of Third Hidden Node (H1_3)

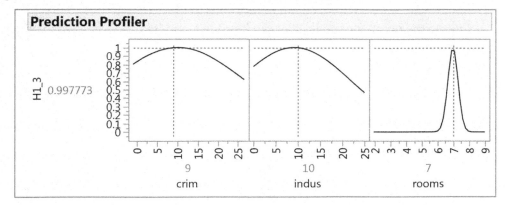

For continuous response neural network models, the prediction equation for the response is a linear combination of the hidden layer outputs. This looks very much like a linear regression model with the hidden nodes as the predictors. The formula for the predicted response, **mvalue**, is shown in Figure 7.12.

Figure 7.12: Prediction Formula for mvalue

$$29.5749097622073 + 11.8994900493049 * H1\_1 + 0.63928019404402 * H1\_2 + -5.3240212964904 * H1\_3$$

The full prediction formula, when written out in expanded form, is complicated and difficult to understand. A profiler view of this prediction formula for **mvalue** is shown in Figure 7.13. We can see that it has combined both the linear and non-linear relationships between the response and inputs that we initially observed (see Figure 7.1).

Figure 7.13: Profiler View of Prediction Formula for mvalue

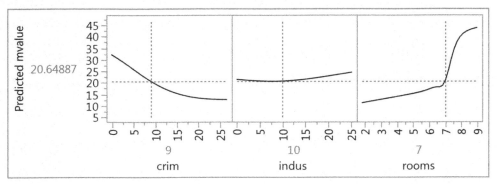

If we wish to make more precise predictions with less residual error, we could add more nodes with our choice of activation functions to the neural network model. There is no limit to the number of nodes that can be added, and it is not necessary to use all three types of activation functions.

JMP Pro also allows you to add a second hidden layer to the model. A more complicated two hidden layer neural network for this example is shown in Figure 7.14.

Figure 7.14: Neural Network with Two Hidden Layers

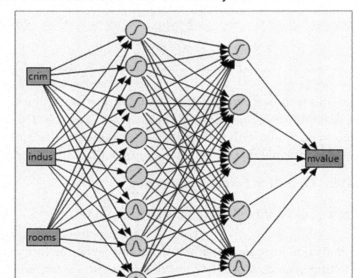

The ability to add a second layer, increase the number of nodes, and apply different activation functions makes neural networks extremely flexible, and the resulting models will generally predict the fitted data very well. This makes neural networks ideal for modeling even the most complex relationships. But, there is a tradeoff: adding model complexity makes the model more difficult to interpret, and more complex models tend to over-fit the data.

A Bit More on the Workings of Neural Networks

The **Neural** platform in JMP Pro, unlike other modeling platforms, requires some form of model cross-validation. JMP Pro uses a portion of the data (either pre-specified or selected at random) to aid in the model-building process and to help prevent over-fitting.

The Neural platform also applies a penalty to prevent over-fitting. This penalty is related to the overall size (magnitude) of the model parameters. In the example model shown in Figure 7.2, there are 16 model parameters corresponding to the weights and intercepts for each of the hidden nodes and the final prediction equations (see Figures 7.4, 7.7, 7.10, and 7.12). The size of the penalty applied to these parameters can be calculated in one of four ways in JMP Pro. "Squared," the default method, is generally the best to use when you believe that most of the selected predictor variables contribute to the predictive ability of the model. The penalty that is applied is the sum of the squared values of the parameter estimates, multiplied by a penalty factor (λ). The optimum value of the penalty factor is determined by examining the model predictions from the holdout portion of the data. This "parameter penalization" is a form of *regularized regression*, which is discussed in more detail in Chapter 9. For more details on the penalty methods available in the Neural platform, see the JMP Help or refer to the book *Specialized Models* (under **Help > Books**).

Neural Networks with Categorical Responses

In the **Boston Housing** example above, we built a neural network to predict a continuous response, but neural networks can also be used with categorical responses. For categorical output variables, instead of predicting the mean, the neural network predicts the probabilities of the outcomes of the categorical response. The neural network structures for continuous and categorical responses are similar, but there is one key difference. For categorical responses, there is an additional logistic transformation that is applied to the final prediction expression. This transformation, which is not shown in the model diagram, is similar to logistic regression described in Chapter 5.

The following examples focus on building and interpreting neural network models. We cover more advanced neural network features and options, including tools that help prevent over-fitting, in Chapters 8 and 9.

Example 1: Churn

Customer retention is a challenge in the ultracompetitive mobile phone industry. A mobile phone company is studying factors related to customer *churn*, a term used for customers who have moved to an alternative service provider. In particular, the company would like to build a model to predict which customers are most likely to move their service to a competitor. This knowledge will be used to identify customers for targeted interventions, with the ultimate goal of reducing churn.

The Data   Churn 2 BBM.jmp

The sample data set consists of 3332 customer records. The response variable of interest is the column called **Churn**, which takes two values:

- **True**: The customer has moved to an alternative service provider.
- **False**: The customer still uses "our" service.

The potential predictors are primarily related to service use and account. A high-level summary of all of the variables, produced with the **Columns Viewer**, is shown in Figure 7.15.

Applying the Business Analytics Process

We start as we always do by getting to know our data. In Figure 7.15, we see that there are three categorical and fifteen continuous factors that can be used as predictors. There are no missing values.

Our response variable, **Churn**, is a two-level categorical variable. In Figure 7.16, we see that 14.5% of the data is for those who have "churned."

Figure 7.15:   Churn Data Summary

Columns	N Categories	Min	Max	Mean	Std Dev	Median	Lower Quartile	Upper Quartile	Interquartile Range
Churn	2	.	.	.	.	.	.	.	.
State	51	.	.	.	.	.	.	.	.
AcctLength	.	1	243	101.056723	39.8253478	101	74	127	53
IntlPlan	2	.	.	.	.	.	.	.	.
VMPlan	2	.	.	.	.	.	.	.	.
NVMailMsgs	.	0	51	8.09393758	13.6872867	0	0	20	20
DayMinutes	.	0	350.8	179.74949	54.455494	179.4	143.625	216.375	72.75
DayCalls	.	0	165	100.432773	20.0714121	101	87	114	27
DayCharge	.	0	59.64	30.5579532	9.25741121	30.5	24.415	36.785	12.37
EveMinutes	.	0	363.7	200.981423	50.7214183	201.4	166.6	235.3	68.7
EveCalls	.	0	170	100.114646	19.9256062	100	87	114	27
EveCharges	.	0	30.91	17.0836315	4.31131144	17.12	14.16	20	5.84
NightMin	.	23.2	395	200.858884	50.5757354	201.15	167	235.3	68.3
NightCalls	.	33	175	100.110444	19.5709101	100	87	113	26
NightCharge	.	1.04	17.77	9.03873349	2.27595824	9.05	7.52	10.59	3.07
IntlMin	.	0	20	10.2373649	2.79225556	10.3	8.5	12.1	3.6
IntlCalls	.	0	20	4.47989196	2.46145017	4	3	6	3
IntlCharge	.	0	5.4	2.76460084	0.75388492	2.78	2.3	3.27	0.97
NCustServiceCalls	.	0	9	1.56302521	1.31565234	1	1	2	1

Figure 7.16:   The Distribution of Churn

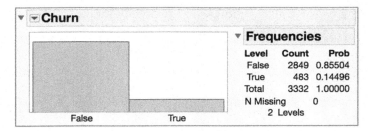

The distributions of some of the potential predictors are shown in Figure 7.17. All of the rows where **Churn=True** are selected in the data table. A potentially good predictor variable is one where the shaded regions in the histogram cluster in one or more regions of the graph.

The length of time someone has been a customer (**AcctLength**) doesn't seem to be related to churn. Those customers who have churned tend to have higher average daytime minute usage and daytime charges, very few voicemail messages, lower than average international calls, and some have more customer service calls than average.

The type of plans that are associated with the account (International - **IntlPlan** or Voicemail - **VMPlan**) look like they may have some relationship with churn, but this is difficult to see from the distributions view.

Figure 7.17: Churn, Distributions of the Potential Predictors

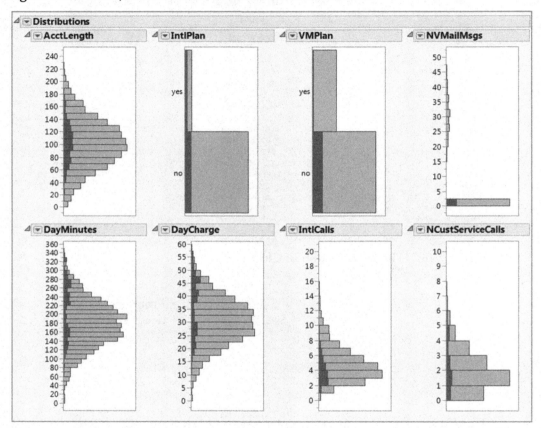

We can examine the relationship between the categorical predictors and churn with a Mosaic Plot (using **Analyze > Fit Y by X** or **Graph > Graph Builder**). The plots of **Churn** versus **IntlPlan** and **VMPlan** are shown in Figure 7.18. Examining these graphs shows that customers with an international plan or without a voicemail plan appear to be more likely to churn.

Figure 7.18:   Churn versus Intlplan and VMPlan

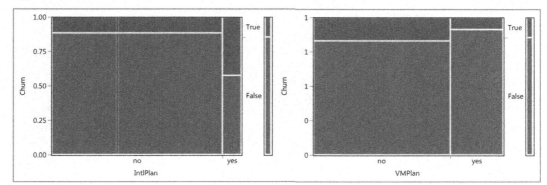

Modeling

After exploring our variables and gaining an understanding of potential relationships, we fit a neural network in JMP Pro using **Analyze > Modeling > Neural** (see Figure 7.19). In this example, we omit the variable **State** and focus on predictors related to call and plan information.

Since we are not missing values from any of the variables, we leave the **Missing Value Coding** box unchecked. When this option is selected, *informative missing coding* of missing values is applied (see "Informative Missing" in Chapter 3).

Figure 7.19: Churn, Neural Dialog Window

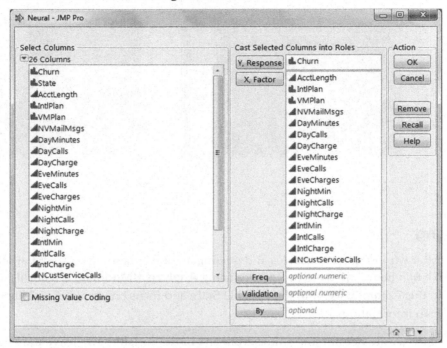

The resulting **Model Launch** dialog (shown in Figure 7.20) lets us specify the structure of the neural network model and other fitting options.

We use the default model that has a single hidden layer with three nodes, each with TanH activation functions. We also use the default **Holdback Validation Method**, using 1/3 of the data as the holdback portion. The rows used for the holdback sample can be saved as a column in the data table (select **Save Validation** from the red triangle for the fitted model).

We use the default fitting options of **Squared Penalty Method**, and do not select the option to **Transform Covariates** (see the JMP Help, or the book *Specialized Models*, for details on this option). The options for **Boosting** will be discussed in Chapter 9.

Figure 7.20: **Churn, Specifying the Neural Network Model**

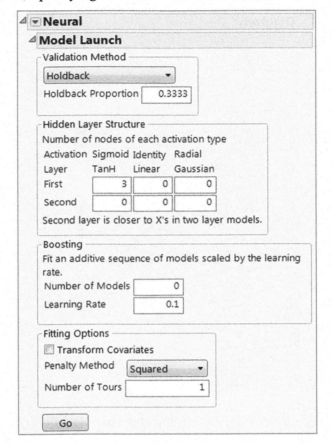

The Neural Model and Results

After running the model (click **Go** at the bottom), we can display the model structure (select **Diagram** from the red triangle for the model). We see the input variables mapping to each of the activation functions in the hidden layer, and the nodes in the hidden layer mapping to the output layer (Figure 7.21).

Figure 7.21:   Churn, Diagram for the Fitted Neural Network Model

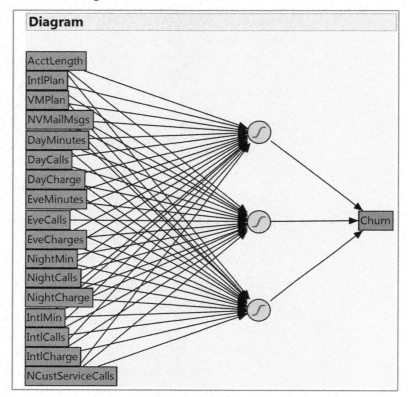

The model results for both the training and validation sets are shown in Figure 7.22.

> *Note: Since a random holdout sample is used, your model results may be slightly different. To obtain the same results shown in Figure 7.22, set the random seed to 1000 prior to launching the **Neural** platform. As a reminder, the random seed can be set with the **Random Seed Reset** add-in, which can be installed from the **JMP File Exchange**.*

The response variable (**Churn**) for this model is categorical. As we've seen with logistic regression and classification trees, the confusion matrix and overall misclassification rate provide indications of the predictive ability of our model. Since the validation set wasn't directly used in estimating the model parameters, it provides a less biased assessment of model performance.

The misclassification rate for the validation data is 9.1%. Examining the confusion matrix we see that for customers who actually churned, 47.2% of the time the model predicted they would indeed churn.

Figure 7.22:   Churn, Neural Network Model Results

◢ ▾ Model NTanH(3)					
◢ Training			**◢ Validation**		
◢ Churn			**◢ Churn**		
Measures		**Value**	**Measures**		**Value**
Generalized RSquare		0.6304189	Generalized RSquare		0.5476305
Entropy RSquare		0.5295728	Entropy RSquare		0.4453264
RMSE		0.2287685	RMSE		0.2554901
Mean Abs Dev		0.1085936	Mean Abs Dev		0.1292547
Misclassification Rate		0.0751914	Misclassification Rate		0.0909091
-LogLikelihood		432.45175	-LogLikelihood		254.9921
Sum Freq		2221	Sum Freq		1111

Confusion Matrix			Confusion Matrix		
Actual	**Predicted**		**Actual**	**Predicted**	
Churn	**False**	**True**	**Churn**	**False**	**True**
False	1877	22	False	934	16
True	145	177	True	85	76

Confusion Rates			Confusion Rates		
Actual	**Predicted Rate**		**Actual**	**Predicted Rate**	
Churn	**False**	**True**	**Churn**	**False**	**True**
False	0.988	0.012	False	0.983	0.017
True	0.450	0.550	True	0.528	0.472

The ROC and Lift Curves (available from the model red triangle) provide additional information about the predictive ability of our model. For example, from the lift curve in Figure 7.23, we see that for the rows that are in the top 20% of the sorted probability of **Churn=True** (Portion = 0.20), there are roughly 4 times more churners than we would expect if we drew 20% of the customers at random.

> *Note: Recall that lift is a comparison of the churn rate for a given portion of the data when sorted in order of predicted probability, compared to the churn rate for the entire population (see the discussion in Chapter 6). We revisit this calculation in an exercise at the end of the chapter.*

Figure 7.23: Churn, ROC, and Lift Curves for the Neural Network Model

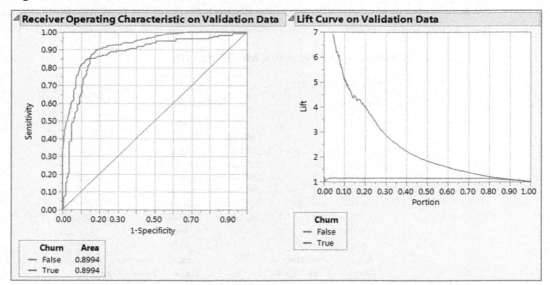

Neural networks have many parameters that must be estimated when building the model. To view these parameter estimates, select **Show Estimates** from the red triangle for the model. For this particular model, there are 58 parameters (54 for the hidden nodes, and 4 for the output prediction equation). These estimates were saved to a JMP data table and rearranged into the format shown in Figure 7.24.

Figure 7.24: Churn, Parameter Estimates for Neural Network Hidden Layer Nodes

Factor	H1_1	H1_2	H1_3
Intercept	-1.524580326	226.78898995	-175.2575364
AcctLength	0.0001851029	0.6602152479	-0.002916288
DayCalls	0.0004026945	1.1245328614	-0.02219087
DayCharge	0.0434450093	-2.653459539	1.6433931857
DayMinutes	-0.008012387	-0.418394044	0.259601226
EveCalls	0.0002851728	1.0555390133	0.0106867044
EveCharges	0.1685751544	-1.579369171	1.3645226852
EveMinutes	-0.014476612	-0.167993951	0.1476986432
IntlCalls	-0.007438036	-3.715040334	0.3649631227
IntlCharge	0.8728402148	-8.788816441	2.6904102381
IntlMin	-0.234597248	-3.469150555	0.0443857527
IntlPlan	1.1062899103	-30.57378459	-73.50036077
NCustServiceCalls	0.0474393156	10.7150937	-0.252640996
NightCalls	0.0002224112	-0.228191976	-0.01716037
NightCharge	0.1666815988	-11.33035307	1.9386737089
NightMin	-0.007654539	-0.518164201	0.0526104332
NVMailMsgs	0.0004526048	0.6151637036	-0.503750598
VMPlan	0.0108520917	-0.922657744	16.418322183

The estimates can also be seen in the saved formulas for the model (from the red triangle for the model, select **Save Formulas**). For this example, six columns will be saved to the data table:

- **Probability(Churn=False)** and **Probability(Churn=True):** These formulas are shown in Figures 7.25 and 7.26.

- Formulas for the three hidden layer nodes (**H1_1, H1_2**, and **H1_3**).

- **Most Likely Churn:** This follows the default decision rule and makes a classification based on which output category (**Churn=True** or **Churn=False**) has the largest predicted probability.

Figure 7.25: Probability(Churn=False) Formula

$$\frac{\text{Exp}\left(-21.495888449054 + -43.807172246693 * H1\_1 + 0.28585263274955 * H1\_2 + -16.246347210095 * H1\_3 \right)}{\left[1 + \text{Exp}\left(-21.495888449054 + -43.807172246693 * H1\_1 + 0.28585263274955 * H1\_2 + -16.246347210095 * H1\_3 \right) \right]}$$

Figure 7.26:   Probability(Churn=True) Formula

$$\frac{1}{\left[1 + \text{Exp}\left[-21.495888449054 + -43.807172246693 * H1\_1 + 0.28585263274955 * H1\_2 + -16.246347210095 * H1\_3\right]\right]}$$

Examining these formulas provides an understanding of the complexity of neural network models and why they can be difficult to interpret. However, the predicted probabilities can be used to better understand the conditions leading to churn. One way to do this is to look at groups of customers with the highest and the lowest probability of churn. This can then be used to establish *profiles* of churners and non-churners, and can help guide marketing campaigns and other customer retention initiatives (Linoff and Berry, 2011).

Another way to get a better understanding of our model is to use the **Categorical Profiler** (from the red triangle for the model). The profiler can be used for making predictions, but also for exploring how the predicted probability of churn changes as the values of the various predictors change. In Figure 7.27, we use the **Arrange in Rows** option (from the red triangle next to **Prediction Profiler**) to display profiles for all of the predictors on one screen.

We see that for customers with international calling plans (drag the line for **IntlPlan** from no to yes), the probability of churning increases as the usage minutes for day, night, evening, and international calls decreases, and also as the charges for those types of calls increase. There appears to be a threshold for each of these factors where churning becomes very likely. Also, the more service calls a customer has had seems to slowly increase the probability of churning. Factors such as the number of calls, account length, voicemail features, and voicemail usage do not seem to be strongly related to the probability of churning.

Figure 7.27:   Churn, Categorical Profiler

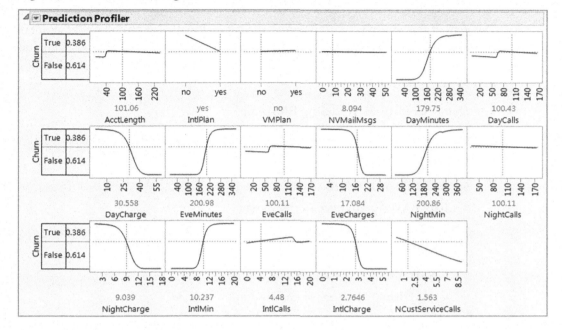

Case Summary

The goal of this study was to build a predictive model for customer churn. We developed a simple neural network, with one hidden layer and three nodes, each using the TanH function, and achieved a misclassification rate for the validation set of 9.1%. We used only one relatively simple neural network model, but performance might be improved with a more complicated neural model. We return to this case study and explore other more complex neural models in an exercise.

Since the target variable in this case study is categorical, we could also have used other modeling methods, such as logistic regression and classification trees. If we fit multiple models, we need to be able to easily compare the predictive performance of these models. In Chapter 8, we show how to compare competing models and identify the "best" model.

We also didn't concern ourselves with variable reduction (choosing only the most important predictors to include in the model) as we did with regression or classification trees. Neural models allow us to use a large number of input variables, even if some of these input variables have a high degree of multicollinearity or have little influence on the response.

Example 2: Credit Risk

The goal of this study is to understand the characteristics of customers who are most likely to default on a loan. The setting is a large bank in the European Union. The target variable is credit risk (**Risk**). This is a 0/1 binary response, where **0 = Good Risk** and **1 = Bad Risk**.

The Data   CreditRiskModeling BBM.jmp

There are twenty-four possible predictors, and 46,500 customers in the data set. The predictors, in alphabetical order, are briefly described below:

AGE: Customer age

BUREAU: Credit bureau risk class

CAR: Type of vehicle

CARDS: Credit card type used by customer

CASH: Requested cash loan amount

CHILDREN: Number of children in household

DIV: Business region

EC_CARD: EC card holder

FINLOAN: Finished paying off previous loans?

INCOME: Personal income level (per week in Euros)

LOANS: Number of loans outside bank

LOCATION: Location of credit bureau

NAT: Nationality

NMBLOAN: Number of loans with bank

PERS_H: Number of persons in household

PRODUCT: Type of credit product

PROF: Profession

REGN: Region

RESID: Residence type

STATUS: Status (internal business metric)

TEL: Telephone

TITLE: Title

TMADD: Time at current address (in months)

TMJOB1: Time at current job (in months)

Applying the Business Analytics Process

Prepare for Modeling

We start by looking at the target variable, **Risk**. The 0/1 response values have been replaced by value labels to make the interpretation easier. As we can see in Figure 7.28, the response is unbalanced, with only a little over 3.2% of the records in the target bad credit risk group.

> With most of the data coming from the **Good Risk** group, we may have a difficult time building a model that predicts the **Bad Risk** outcome, which is the desired result of our modeling work. While we don't directly address this issue in this example, there are strategies that can be employed to deal with the imbalance in the response categories (see Shmueli et al., 2010, and Linoff and Berry, 2011). The simplest approach is to create a modeling data set that uses all of the **Bad Risk** data, but only a smaller random sample of the **Good Risk** data. This undersampling approach creates a modeling data set that has better balance, but care should be taken to fully explore the new modeling data set to ensure that variables in the undersampled **Good Risk** group have similar distributions and relationships when compared to the complete data set.

Figure 7.28: Distribution of Risk

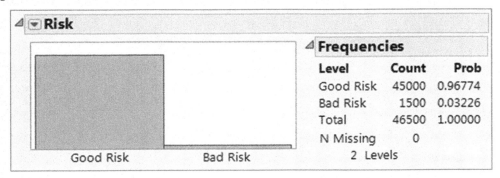

A look at the summary statistics for all of the variables from the **Columns Viewer** (Figure 7.29) highlights some potential data quality issues:

- Missing values for two of the variables, **PRODUCT** and **RESID**.

- Several categorical variables have a continuous modeling type. For example, **FINLOAN**, **DIV**, and **EC_CARD** all have a minimum value of 0 and a maximum value of 1 rather than listing the number of categories.

- Two variables (**TMADD** and **TMJOB1**) appear to have a missing value code of 999.

- **INCOME**, which is the weekly income in Euros, and **CASH**, the requested amount of the loan, both have a minimum value of 0 and a maximum value of 100,000. This may indicate some sort of truncation in the data. For example, incomes and balances > 100000 might be set to 100000.

A look at the distributions of each of the variables (not shown here) reinforces concerns regarding data quality.

Figure 7.29:   Summary Statistics for Credit Risk Variables

Columns	N	N Missing	N Categories	Min	Max	Mean	Std Dev
Risk	46500	0	2	.	.	.	.
TITLE	46500	0	2	.	.	.	.
CHILDREN	46500	0	.	0	23	0.9023010753	1.183075575
PERS_H	46500	0	.	1	25	2.5401505376	1.437666163
AGE	46500	0	.	18	71	36.868774194	11.376900474
TMADD	46500	0	.	0	999	128.60129032	193.51533651
TMJOB1	46500	0	.	0	999	97.209870968	140.12012608
TEL	46500	0	.	0	2	1.8608387097	0.3461806083
NMBLOAN	46500	0	.	0	2	0.6442795699	0.9222678796
FINLOAN	46500	0	.	0	1	0.5066236559	0.4999615012
INCOME	46500	0	.	0	100000	1875.7483871	1652.6149536
EC_CARD	46500	0	.	0	1	0.3347311828	0.4719014803
STATUS	46500	0	6	.	.	.	.
BUREAU	46500	0	.	1	3	1.7051612903	0.9554222635
LOCATION	46500	0	.	0	1	0.9986451613	0.0367835863
LOANS	46500	0	.	0	9	0.9680860215	1.0660930199
REGN	46500	0	.	0	9	3.1929462366	2.5322661819
DIV	46500	0	.	0	1	0.717311828	0.4503109262
CASH	46500	0	.	0	100000	2591.6172043	6168.955835
PRODUCT	46256	244	6	.	.	.	.
RESID	38280	8220	2	.	.	.	.
NAT	46500	0	8	.	.	.	.
PROF	46499	1	9	.	.	.	.
CAR	46500	0	3	.	.	.	.
CARDS	46500	0	7	.	.	.	.

Before developing a model, we need to address these issues if possible. We change the modeling types to nominal for numeric variables that are actually categorical, and set '999' as a missing value code (using **Column Properties > Missing Value Codes**) for **TMADD** and **TMJOB1**. This tells JMP to treat the value 999 for these variables as though it is missing. Because we have missing values, we will use the **Neural** platform option for **Missing Value Coding** when we create our neural model.

Next, we look at the distribution of the target category, **Bad Risk**, across the predictors. Distributions of six of the predictors, with the bad risk observations selected, are shown in Figure 7.30. The imbalance in the response variable makes it very difficult to explore potential relationships between the bad risk level and the predictors, so not much can be learned from this.

An alternative approach for exploring the relationships between the predictors and the response, **Risk**, is to create a tabular summary of the data using **Analyze > Tabulate**. Data filtering, introduced in Chapter 3, is another approach (select **Script > Local Data Filter** from the top red triangle and choose **Risk** as the filtering variable).

Figure 7.30:   Credit Risk, Response Variable GB with Potential Predictors

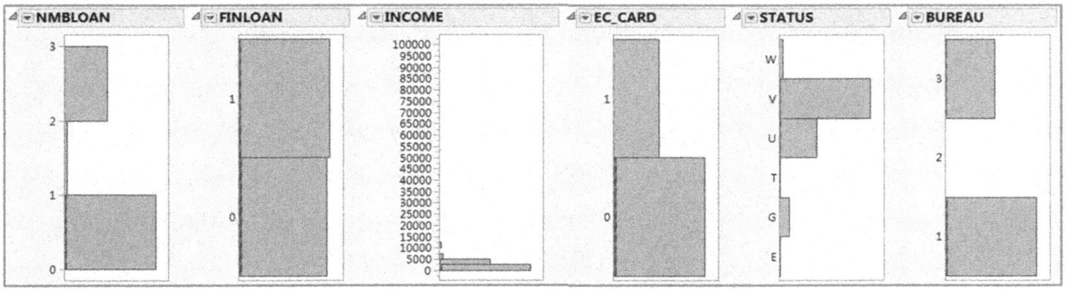

Building the Model

We build a neural network for **Risk** with all 24 potential predictors and check the **Missing Value Coding** box. Again, to produce the same results, use the **Random Seed Reset** add-in to set the random seed to 1000 before launching the **Neural** platform.

For this example, we use two hidden layers. The hidden layers have three nodes, and each node applies a different activation function. We use the default **Holdback** validation method, with a **Holdback Proportion** of 0.3333, and accept the remaining default settings. Note that this may take a couple of minutes to run.

Figure 7.31: Credit Risk, Neural Network Model Launch

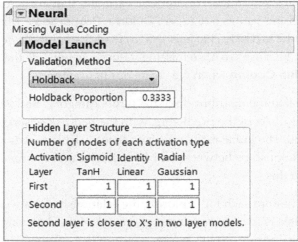

The model diagram is shown in Figure 7.32. Note that the second hidden layer is closest to the inputs, and the first hidden layer is closest to the output.

Figure 7.32: Credit Risk, Neural Network Model Diagram

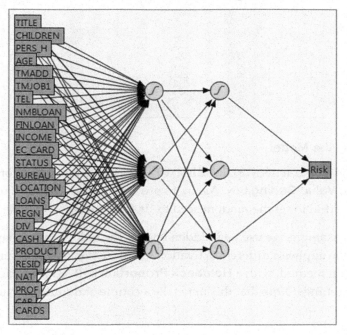

Evaluate the Model and Interpret Results

The misclassification rate for both the training and validation sets are around 3% (see Figure 7.33). In the validation set, we correctly classify all of the **Good Risk** customers, but only 22 of the 500 customers in the **Bad Risk** category were classified by the model as being bad risks. This is due, in part, to the imbalance in the response variable and the fact that the standard threshold used for classification is a probability of 0.5.

Figure 7.33: Credit Risk, Neural Network Model Results

Model NTanH(1)NLinear(1)NGaussian(1)NTanH2(1)NLinear2(1)NGaussian2(1)

Training

Risk

Measures	Value
Generalized RSquare	0.1892528
Entropy RSquare	0.1686629
RMSE	0.1649675
Mean Abs Dev	0.0552513
Misclassification Rate	0.0294184
-LogLikelihood	3672.6101
Sum Freq	31001

Confusion Matrix

Actual Risk	Predicted Good Risk	Bad Risk
Good Risk	30001	0
Bad Risk	912	88

Confusion Rates

Actual Risk	Predicted Rate Good Risk	Bad Risk
Good Risk	1.000	0.000
Bad Risk	0.912	0.088

Validation

Risk

Measures	Value
Generalized RSquare	0.1325538
Entropy RSquare	0.1172754
RMSE	0.1693755
Mean Abs Dev	0.056757
Misclassification Rate	0.0308407
-LogLikelihood	1949.7694
Sum Freq	15499

Confusion Matrix

Actual Risk	Predicted Good Risk	Bad Risk
Good Risk	14999	0
Bad Risk	478	22

Confusion Rates

Actual Risk	Predicted Rate Good Risk	Bad Risk
Good Risk	1.000	0.000
Bad Risk	0.956	0.044

In this situation, the Lift Curve and the ROC Curve can give a better indication of the ability of our model to correctly classify bad risks.

The Lift Curve is shown on the right in Figure 7.34. The lift in the top portion of our sorted probabilities for bad risk (the blue line) starts high, drops quickly, and then tapers off. In the top 5% (portion = 0.05), the lift is roughly 5. This means that there are 5 times as many bad risk customers in the top 5% of the sorted probabilities of bad risk than we would expect to see if we just chose 5% of the data at random.

The ROC Curve (on the left in Figure 7.34) provides assurance that our model performs better than a random sorting model. The area under the curve (AUC) for **Bad Risk** is 0.74. (Recall that the AUC for a random sorting model is 0.5, and the AUC for a perfect sorting model is 1.0.)

Figure 7.34:   Credit Risk, Lift, and ROC Curves for Risk

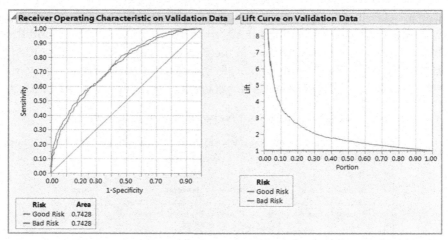

As we discussed in the first example, the **Categorical Profiler** can help us get a sense of the relationships between the predictor variables and the target variable. In Figure 7.35, the value for **AGE** has been changed to **25**, and the level for **CARDS** is set to **VISA mybank**. The contour of the line for each predictor shows what happens to the predicted risk if values of the predictor are changed.

What would happen to the probability of **Bad Risk**, with all else held constant, if the value for **LOANS** is increased?

Figure 7.35:   Credit Risk, Neural Network Model Prediction Profiler

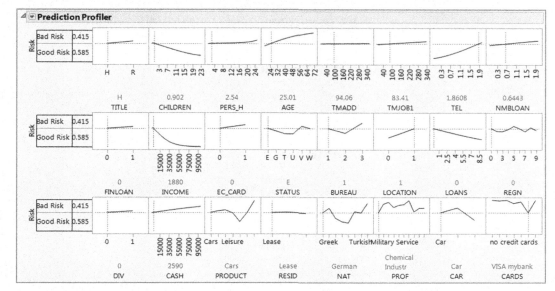

To help us sort through the large set of potentially important variables, we use the **Variable Importance** option. This is available under the red triangle for the **Prediction Profiler** (select **Assess Variable Importance > Independent Uniform Inputs**). This uses simulation to estimate the sensitivity of the response to changes in each of the predictor variables (see the **JMP Help** for details).

The **Variable Importance Summary Report** provides information similar to the column contributions report in the **Partition** platform. We see that **INCOME** and **CARDS** are the two most important variables (Figure 7.36) and that several of the predictors are not very useful in predicting **Risk**.

The **Marginal Model Plots** show the factors, ordered by the total effect in the **Variable Importance Summary Report** (see Figure 7.37). The contours in the plot show the average response for each predictor across the distributions of the other predictors.

Figure 7.36:   Credit Risk, Variable Importance

Column	Main Effect	Total Effect	.2 .4 .6 .8
INCOME	0.128	0.185	
CARDS	0.075	0.124	
PRODUCT	0.047	0.08	
PROF	0.03	0.062	
NAT	0.026	0.055	
LOCATION	0.016	0.036	
TEL	0.018	0.033	
AGE	0.018	0.032	
CAR	0.016	0.028	
CHILDREN	0.012	0.022	
BUREAU	0.008	0.02	
LOANS	0.009	0.014	
STATUS	0.004	0.013	
CASH	5e-4	0.009	
REGN	0.002	0.008	
PERS_H	0.002	0.007	
NMBLOAN	0.002	0.006	
EC_CARD	0.003	0.006	
TITLE	0.002	0.005	
DIV	0.001	0.004	
FINLOAN	0	0.004	
TMJOB1	2e-4	0.003	
RESID	0.001	0.003	
TMADD	0.001	0.002	

Variable Importance: Independent Uniform Inputs
Summary Report

Figure 7.37:   Credit Risk, Marginal Model Plot of the Top 9 Effects

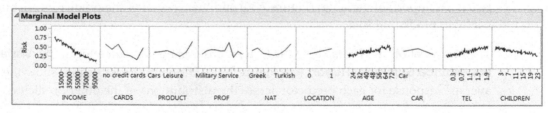

Case Summary

Neural networks are very powerful and flexible prediction models. For this example, the additional tools in JMP for visualizing models and exploring variable importance have helped us uncover important factors that influence credit risk. Using these factors, we can inform other business decisions or identify procedural changes, such as:

- implementing new marketing strategies for identifying customers with low risk of not paying off their loans,

- implementing a proactive or predictive approach to identifying customers that are likely to default, or

- using the model that was developed, we could create a monthly report that ranks each customer according to their default risk, and additional interventions could be made to help the higher risk customers manage their loan payments and finances.

Summary and Key Take-Aways

Neural networks are very useful models when the main goal is prediction. These models can be difficult to interpret if we rely simply on the prediction functions that are created. With the right analytical tools to help you interpret the models, you can also use them to gain insight into the relationships between the inputs and output of the model.

Overfitting is also a concern with neural models. Because of this, JMP requires the use of model cross-validation to help prevent the model from overfitting. See Chapter 8 for a more thorough discussion of cross-validation.

In this chapter, we showed the results of building just one neural network model in each example. In practice, we generally build many models, and explore the impact of adding additional layers and different activation functions in order to find a "best performing" model. The **Boosting** option of the **Neural Model Launch**, which was not discussed in this chapter, provides another approach to choosing the best structure for the model (see Chapter 9).

Exercises

Exercise 7.1: Use the **Churn 2 BBM.jmp** data set for this exercise. In the example, we fit a neural network with one hidden layer and three nodes, using the sigmoid, linear, and Gaussian activation functions.

    a.   Fit the model described in Example 1, and also fit a model with one hidden layer with three nodes, but only use the sigmoid activation function. Save these formulas to the data table. How does this model compare to the original model in terms of misclassification?

    b.   Fit a final model with two hidden layers and several nodes. How does this model compare to the two single-layer models fit in part (a)?

    c.   For the model described in Example 1, compute the lift at portion = 0.2, as described in Chapter 6. What is the overall churn rate? What is the churn rate for the top 20% of the probabilities?

    As a reminder, to compute lift:

    1.   Select **Save Formulas** from the red triangle for the model.

    2.   In the data table, sort the **Probability(Churn=True)** column in descending order (right-click on the column heading and select **Sort**, then **Descending**).

    3.   Calculate the churn rate for the entire data set (see Figure 7.16).

    4.   Calculate the churn rate for the top 20% of the probabilities.

    5.   Divide the results from #4 by the results from #3. This is the lift at portion (percentile) = 0.2.

    d.   Manually compute the lift at portion = 0.1, as you did in part c.

    e.   Recall that this team is tasked with identifying customers most likely to churn, with the goal of developing strategies or interventions to minimize the risk of churn. How can this model be used toward this goal?

Exercise 7.2: Use the **Boston Housing BBM NN.jmp** data set for this exercise. Fit a neural network with the continuous variable **mvalue** as the response and the other variables as the predictors. Use the default settings.

    a.   What is the validation RSquare and the RMSE?

    b.   Display the actual by predicted plots. Are there any unusual patterns or observations?

    c.   Use the prediction profiler to explore relationships between the predictors and the response. Which combination of factor settings leads to the highest predicted **mvalue** (median value)?

    d.   Open the variable importance report. Which variables are most important? Describe the nature of the relationship between the top three variables and the response (i.e., what happens to the predicted response as the value of each predictor is changed)?

Exercise 7.3: Use the **CreditRiskModeling BBM.jmp** data for this exercise. In this chapter, we fit a model with 2 hidden layers.

    a.   Fit a single-layer model with one node using the TanH function. Compare the results of your model to those of the model created earlier in this chapter.

        1.   Which model is better in terms of model misclassification and the confusion matrix?

        2.   Is there a difference in the Lift Curve or the ROC Curve? If so, which model performs better?

        3.   Which model would you rather explain to a manager? Why?

        4.   Here's a scenario: You've built a neural network model to predict credit risk and have identified variables related to bad risk. You need to communicate your findings and your conclusions to your manager. Prepare a short (3-5 minute) report. Include a summary of your objectives, the data used, the model and key output from your analysis, and your conclusions.

    b.   Now, fit a two-layer model with several nodes and different activation functions.

        1.   How does this model compare to the other two models?

        2.   Which model is the best? Why?

Exercise 7.4: Use the **CreditRiskModeling BBM.jmp** data for this exercise. In the chapter, we saw that the maximum value for the variable **INCOME** is 100000.

    a.   How many customers reported a weekly income of 100000? Is it likely that this is a valid number?

    b.   For this exercise, let's assume that 100000 is actually a missing value. Use the **Missing Value Code** column property for **INCOME** and 100000 as a missing value code. Use **Columns Viewer** or **Distribution** to verify that 100000 no longer appears, and that there is at least 1 missing value for **INCOME**.

    c.   We want to compare models fit with and without this extreme **INCOME** value. First, set the random seed to 1000 (as shown earlier in this chapter). This will ensure that the model is built on the same 2/3 of the data. Then, fit the same neural model illustrated in this chapter for credit risk.

    d.   Are there any differences in the model statistics, the Lift Curve, or the ROC Curve?

    e.   Open the **Categorical Profiler** and the **Variable Importance** report. Does INCOME still show as the most important variable?

Exercise 7.5: Again, we use the **CreditRiskModeling BBM.jmp** data. In the chapter, we used variable importance to identify the most important variables.

    a.   Build a neural network model using only the top 3 or 4 most important variables. (Use the default settings, and again set the random seed first.)

    b.   Compare the fit statistics, Lift, and ROC from this model to the model in the chapter.

    c.   Which model is better? Why?

    d.   Which model would you rather explain to your manager?

References

SAS Institute Inc. 2015. JMP 12 *Specialized Models*. Cary, NC: SAS Institute Inc.

Linoff, Gordon S. and Berry, Michael J. A., 2011. Data Mining Techniques, 3rd ed. Chapter 2, page 41, Wiley.

Shmueli, Galit, Nitin R. Patel, and Peter C. Bruce. 2010. *Data Mining for Business Intelligence: Concepts, Techniques, and Applications in Microsoft Ofice Excel with XLMiner*, 2nd ed. John Wiley & Sons, Inc.

Part 4

Model Selection and Advanced Methods

Part IV introduce methods for validation of predictive models (**Cross-Validation**), and with this in mind revisits examples introduced in previous chapters. In **Advanced Methods** we introduce some advanced model-building tools and techniques. Last, we conclude with **Capstone and New Case Studies.** We revisit the Business Analytics Process, applying the entire process to a new case study, and introduce new examples based on large and messy data sets that are more representative of real business problems.

8

Using Cross-Validation

Overview

We first introduced model cross-validation in Chapter 6. In this chapter, we dig more deeply into the concept of cross-validation and describe commonly used cross-validation methods that are available in JMP Pro. We show how to use cross-validation to aid in choosing the best model, and see how to compare and evaluate different models using the JMP Pro **Model Comparison** platform.

Why Cross-Validation?

A famous expression attributed to the imminent statistician George Box is "essentially, all models are wrong, but some are useful" (Box and Draper, 1987). Useful models give us accurate insights into the relationships between the inputs and outputs of the process and allow us, for a given set of inputs, to make good predictions about the output. However, according to George Box, any statistical model that we choose to describe the behavior of a process or to make predictions about a process is, at best, an approximation.

Consider the data shown in Figure 8.1. This data set was simulated with an underlying relationship between Y and X, but there is also random measurement error in the Y variable. The real model that describes this data is unknown to the data analyst, but linear regression models can be used to find a good "approximate" model that will allow for good prediction. Figure 8.1 shows the fit of linear, cubic, and 6th-order polynomial models to this data.

Figure 8.1: Competing Models

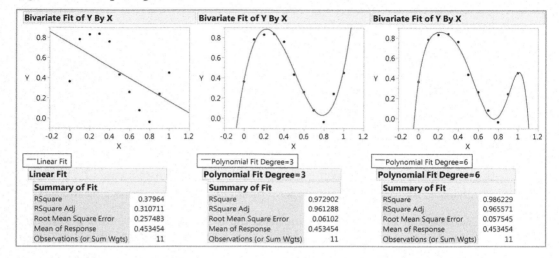

Each model is progressively more complex. The linear model (left) has an RSquare value of 0.38, the cubic model (middle) has an RSquare value of 0.97, and the 6th-order polynomial model (right) has an RSquare value of 0.98.

The fitted equations for the three models are:

$$\text{Model 1: } \hat{Y} = 0.742 - 0.576\,X$$
$$\text{Model 2: } \hat{Y} = 0.370 + 5.08\,X - 14.6\,X^2 + 9.59\,X^3$$
$$\text{Model 3: } \hat{Y} = 0.363 + 7.09\,X - 41.8\,X^2 + 137\,X^3 - 261\,X^4 + 245\,X^5 - 85.2\,X^6$$

Recall that RSquare (also written as R^2) is a measure of how well a model fits the data. When the response is continuous, as it is in this example, then RSquare is defined as:

$$R^2 = 1 - \frac{SSE}{SST}$$

where SSE is the sum of the squared prediction errors (sum of the squared residuals) for the model, and SST is the sum of squares "total" for the response. If we let Y_i be the actual response value for each observation, let $\hat{Y}_i$ be the fitted or predicted value for each observation, and let N be the number of rows in the data set, then:

$$SSE = \sum_{i=1}^{N} (Y_i - \hat{Y}_i)^2$$

and

$$SST = \sum_{i=1}^{N} (Y_i - \overline{Y})^2 .$$

RSquare can take on values between 0 and 1. An RSquare of 1 means that the model predicts perfectly (there is no prediction error, SSE = 0), and an RSquare of 0 means that the model doesn't do any better than just using the response mean, $\overline{Y}$, to make predictions. RSquare is a good general metric to use when comparing models.

We can see that as we add more model complexity (in this case, more polynomial terms) in the example above, we get a better fit to the data. Based on the RSquare for each fitted model, we might choose the 6th-degree polynomial as the "best" model.

But is this a useful model? Using RSquare does have a limitation: As you add more model complexity, in general, RSquare will increase, even if that added model complexity isn't really necessary. Determining how much complexity is needed in a model is an important part of the model-building process.

If the goal of building the model is to be able to make accurate predictions in the future for any value of X, then it is reasonable to check to see how a candidate model would predict data that wasn't used to fit the model. In this example, we collect a *validation* data set of 11 additional points. These points are shown in Figure 8.2, along with the original data used to fit the models (which we call the *training* set).

Figure 8.2: Training and Validation Data

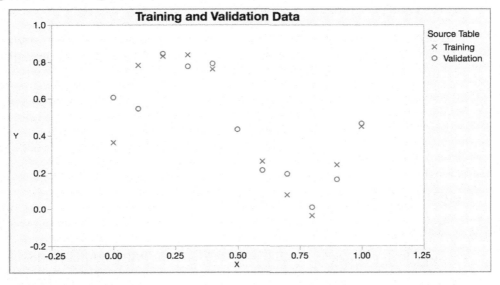

We use the fitted models above to predict these additional data points, and we compare the RSquare on just the validation data set for each of the three models (these are the last three rows in Figure 8.3). Based on the validation RSquare, the cubic model appears to be the best model.

This model also has the lowest *RASE* (Root Average Squared Error) and *AAE* (Average Absolute Error), which are both measures of overall prediction error. For both of these measures, lower values are preferred.

Figure 8.3: Training and Validation RSquare for Each Model

▼ **Measures of Fit for Y**

Source Table	Predictor	Creator	.2 .4 .6 .8	RSquare	RASE	AAE	Freq
Polynomial Models Training Data	Predicted Y (6th Order)	Bivariate Polynomial Fit Degree=6		0.9862	0.0347	0.0276	11
Polynomial Models Training Data	Predicted Y (Cubic)	Bivariate Polynomial Fit Degree=3		0.9729	0.0487	0.0391	11
Polynomial Models Training Data	Predicted Y (Linear)	Bivariate Linear Fit		0.3796	0.2329	0.2029	11
Polynomial Models Validation Data	Predicted Y (6th Order)	Bivariate Polynomial Fit Degree=6		0.8130	0.1173	0.0853	11
Polynomial Models Validation Data	Predicted Y (Cubic)	Bivariate Polynomial Fit Degree=3		0.8429	0.1075	0.0807	11
Polynomial Models Validation Data	Predicted Y (Linear)	Bivariate Linear Fit		0.4727	0.1969	0.1779	11

The fact that the 6th-order polynomial fits the training data set better than it does the validation data can be attributed to *over-fitting* the data. Over-fitting can occur when the model is so complex that it attributes the random noise in the response data to the factors being used to make predictions. Without the use of external validation data, it is difficult to detect that over-fitting is occurring.

This process of using external data for determining the necessary model complexity is what we refer to as *cross-validation*. Many of the modeling platforms in JMP Pro enable you to easily use cross-validation as part of the model fitting process.

> **Note**: *In many textbooks and software packages, separate terms such as "validation," "hold-out validation," "test/train/validate," and "k-fold cross-validation" are used. We group all of these approaches together under one general descriptive term, "cross-validation."*

Partitioning Data for Cross-Validation

Using a Random Validation Portion

One way to implement cross-validation is to have the JMP platform randomly identify a specified proportion or percent of the rows to be used for validation. The remaining rows are then used to train (build) the model. Many JMP platforms have this capability. For example, as we saw in Chapter 6, when using the **Partition** platform, there is an option to specify a **Validation Portion** (left, in Figure 8.4), which is the proportion (a value between 0 and 1) of the data that will be chosen, at random, to be the validation set.

Figure 8.4: Decision Tree Validation Portion Option

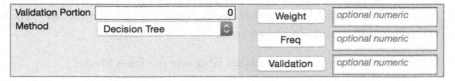

The advantage of using a hold-out portion is that it is quick and easy. Some cautions to consider, however, when using this approach:

- The random selection of a validation portion may not necessarily create a good validation subset. This can occur just due to random chance, or if the proportion of the data selected for validation is too small. To overcome this issue, we recommend that you start by using a validation portion in the range 0.25 to 0.50, so that 25%-50% of your data is being used for validation. Keep in mind, however, that it is difficult to give a firm rule for the percentage of data to use for validation because this will depend on the particular data set. Because of this, we recommend repeating the model fitting process with different randomly held-out validation portions so that you can examine the stability of the models developed with the different validation subsets.

- Since the validation set is chosen at random, using a validation portion doesn't allow you, in general, to reproduce the same model if you fit the model again. In some (but not all) of the JMP Pro fitting platforms, when you request a random validation portion in the launch or model fitting dialog, you can save an indicator column of the **Validation Role** for each row in the data table. This column can be used again later to re-create the model fit with the same test and validation subsets.

    Note: Another alternative is to use the **Random Seed Reset** *add-in (or the JMP Scripting Language) to specify the starting random seed prior to fitting the model. This same random seed can be used to re-create the model fit if needed. In practice, setting the random seed isn't recommended, but for the purposes of teaching and training, this can be useful to create repeatable results.*

Specifying the Validation Roles for Each Row

A very common method of cross-validation is to divide the data set into *training*, *validation*, and (optionally) a *test* subset. This is done by creating an indicator column in the data table to specify the validation role for each row in the data table. The three roles are specified by a numeric column, typically with the values 0 (for training), 1 (for validation) and 2 (for test). How these subsets are used in model building, evaluation and selection is described below:

- The *training* subset is used to fit the statistical model(s), and the coefficients for terms in the model are calculated using only this subset.

- The *validation* subset is used to determine how much complexity is needed in the statistical model. In the model-building process, there can be a number of choices made about model complexity, including:
    - which terms to include in the model (stepwise or generalized regression),
    - the number of branches or nodes in a recursive partitioning model (decision tree, bootstrap forest), and
    - the number of layers of additive models to be used (boosted tree or boosted neural network).

For every decision that has to be made about model complexity, the ability of the model to predict is assessed using the validation set. The model-building process, then, is guided by the performance of the model on the validation subset. The model that maximizes RSquare (or a similar related measure) *on the validation subset* is chosen.

- The *test* subset is optional, but is very much recommended when using cross-validation. This subset is completely held out of the model building process, and has no influence on the fitted model parameters nor on the model complexity. Evaluating the model on a test subset gives an unbiased assessment of how the model would perform when making predictions on additional cases that the model has never seen. This "fair" or "honest" assessment is useful in estimating how well a model will perform for the business problem that it is being developed to solve. Also, if a choice must be made between models that have been built for a given business problem, the models can be evaluated using this external data subset, providing a good side-by-side comparison of the competing models.

K-fold Cross-Validation

A shortcoming of using hold-out samples is that the random selection of a test and validation subsets could result in a poor model, entirely by chance. Poorly chosen hold-out subsets can lead to models that are very biased in their predictions and as a result do not perform well. Also, in some situations, there may not be enough data to create completely separate hold-out data subsets. A way to deal with this problem is to use *k-fold cross-validation*.

During the model building process, every time a decision is made about adding more model complexity, the entire data set is divided into k subsets (also known as *folds*). One of the folds is held out as a validation subset while the other k-1 folds are used to train the model. This process rotates so that each fold has the opportunity to be the validation subset. After all iterations of rotating through the combination of k folds, an RSquare is calculated for the k validation folds. These k RSquare values are averaged to estimate the model validation RSquare. See Figure 8.5 for a graphical representation of this process of determining the RSquare value using k-fold cross-validation.

Figure 8.5: K-fold Cross-Validation

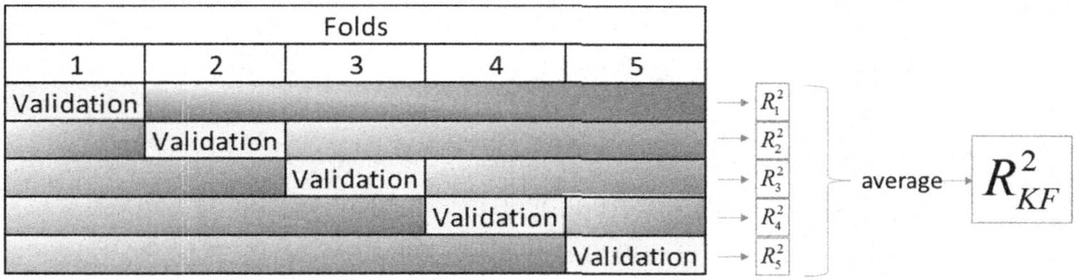

K-fold cross-validation, for the modeling platforms where it is available, is selected as an option in the model fitting dialog or command.

Using Cross-Validation for Model Fitting in JMP Pro

Cross-validation can be used in many of the JMP Pro modeling platforms. Table 8.1 lists the types of cross-validation available in the modeling platforms covered in this book.

> **Note**: *This list is current as of JMP 12. Consult the JMP documentation for more details and the most up-to-date information on this topic.*

Table 8.1

Platform	Use Excluded Rows as Validation Holdback	Random Validation Holdback	K-fold Cross-Validation	Validation Role Column
Fit Least Squares	No	No	No	Yes (only for model evaluation)
Stepwise Regression (Fit Model)	No	No	Yes (continuous response models only)	Yes
Logistic Regression	No	No	No	Yes (only for model evaluation)
Decision Tree	Yes	Yes	Yes	Yes
Bootstrap Forest	Yes	Yes	No	Yes
Boosted Tree	Yes	Yes	No	Yes
Neural	Yes	Yes	Yes	Yes
Generalized Regression	Yes	Yes	Yes	Yes

It is important to remember that the specific way that cross-validation is used may differ between modeling platforms. For example:

- In forward stepwise regression, cross-validation is used to decide when to stop adding additional terms into the model.

- In Neural, cross-validation is used to determine the best value for the penalty parameter used to prevent over-fitting of the neural network.

For more details on how validation is applied in the specific JMP Pro modeling platforms, refer to *Fitting Linear Models* or *Specialized Models* under **Help > Books**.

Example

We use the Boston Housing data set, which was briefly introduced in Chapters 3 and 7, to illustrate how cross-validation is used in building a decision tree model.

The Data   Boston Housing BBM Ch8.jmp

As a reminder, the objective is to use this data set to develop a model to predict the median value of homes in the Boston area. The data were originally collected and assembled in mid 1970s (Harrison and Rubinfield, 1978), so this example is a bit dated. However, it is typical of a socioeconomic data set that is used to inform economic or public policy decisions. Each row in the data table is from a census tract (or town) in the Boston area.

The variables in the data set are:

mvalue: Median value of homes in the census tract

crim: Per capita crime rate for the census tract

zn: Proportion of a town's residential area zoned for lots larger than 25000 square feet

indus: Proportion of non-retail business acres per town

chas: This is a 0/1 indicator variable. If the town bounds on the Charles River, the value is 0; otherwise the value is 1.

nox: Average annual nitrogen oxide concentrations in parts per hundred million

rooms: Average number of rooms in owner-occupied units

age: Proportion of owner units built prior to 1940

distance: Weighted distances to five employment centers in the Boston region

radial: Index of accessibility to radial highways

tax: Full value property tax rate (per $10,000).

pt: Pupil-to-teacher ratio by town school district

lstat: Proportion of population that is "lower status," that is, proportion of adults without some high school education or that are classified as laborers

Creating Training, Validation, and Test Subsets

JMP Pro provides several ways to automatically create a validation column. An easy method is to use the **Make Validation Column** utility (under **Cols > Model Utilities**). In the **Make Validation Column** dialog in Figure 8.6, we request a 50% training, 25% validation, and 25% test split. In this example, we use the **Purely Random** method to create the holdback sets. The name of the validation column that we create is **Validation1**.

> *Note: Alternatively, **Stratified Random** will perform the random selection with balance across specified variables. For instance, if we want each validation subset to contain the same relative proportions of **chas** = **0** and **chas** = **1**, we could use **chas** as the stratification variable.*

Figure 8.6:   Using the Make Validation Column Utility

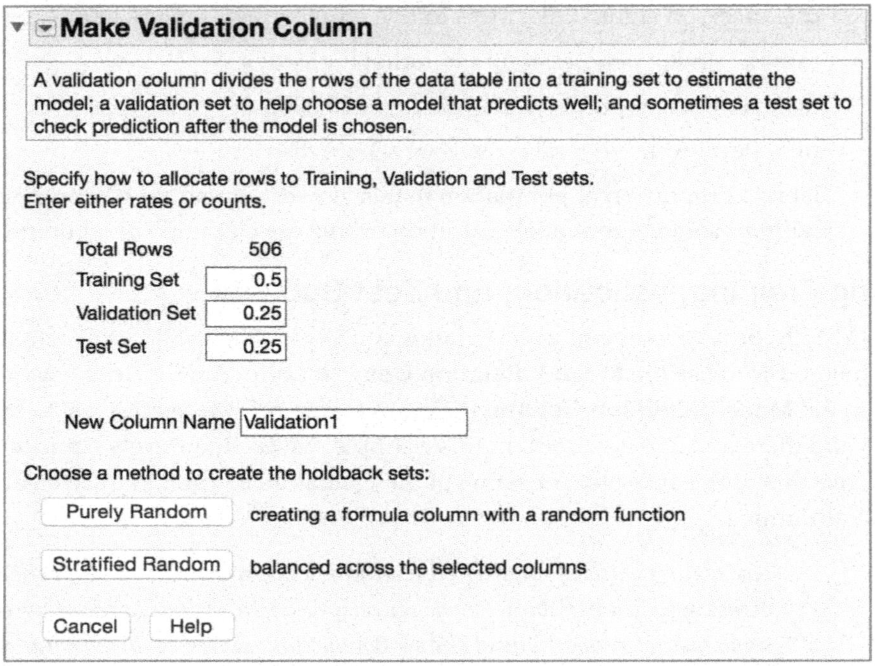

A new column, **Validation1**, is added to the data table (Figure 8.7). The values in this new column are displayed as **Training**, **Validation**, and **Test**. The modeling platforms in JMP require that the validation set indicators be numeric, so the actual values stored are the numbers **0, 1**, and **2**.   The utility also creates a **Value Labels** column property to display descriptive labels in place of the numeric values in the data table and output reports.

Figure 8.7: Validation Column Added to Data Table (First 12 Rows)

mvalue	Validation1
24	Validation
21.6	Training
34.7	Training
33.4	Validation
36.2	Training
28.7	Test
22.9	Training
27.1	Training
16.5	Validation
18.9	Validation
15	Training
18.9	Training

Value Labels

If a column has value labels, and Use Value Labels is checked, the labels are displayed wherever the column data are displayed.

0 = Training
1 = Validation
2 = Test
optional item

[Add] [Change] [Remove]

☐ Allow Ranges

Value [   ]

Label [   ]

☑ Use Value Labels

Examining the Validation Subsets

Because we are simply choosing the validation subsets at random, it is a good idea to check if the random assignment of rows to the training, validation, and test sets resulted in subsets that are reasonably similar to each other. This is mostly a subjective and graphical assessment, but it is important. If the subsets are markedly different, it can lead to unexpected results from the modeling process.

First, we use the **Fit Y By X** platform, with **mvalue** as the **Y, Response** and **Validation1** as the **X, Factor** to examine the distribution of response variables across each of the validation subsets (Figure 8.8). Here, we can use a variety of graphical and numeric displays, such as quantiles and box plots, mean and standard deviations, and density plots (all available from the red triangle) to examine the similarity of the distributions of **mvalue** across the validation roles.

Figure 8.8: Fit Y by X, mvalue by Validation1

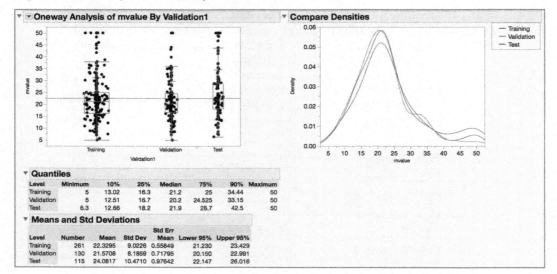

We repeat this process for each of the continuous predictors, to assess whether the distributions of the predictor variables are similar across each of the validation subsets (not shown).

Next, we examine the relationship between the categorical predictor **chas** and **Validation1**. Again, we use **Fit Y by X**, with **chas** as the **Y, Response** and **Validation1** as the **X, Factor**. This creates a **Contingency Analysis**, with a mosaic plot and a contingency table (see Figure 8.9), which we simplify to display only the Row % (we use the red triangle for the **Contingency Table** to turn other options off). What we see is reasonable balance across the validation roles, with roughly the same percent **chas** = 0 and **chas** = 1 in each of the validation subsets.

Figure 8.9: Fit Y by X, chas by Validation1

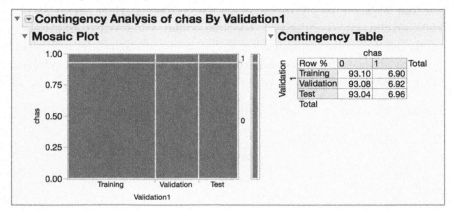

We also examine the simple two-way relationship between the response, **mvalue**, and each of the factors, across the different training sets. This is easy to do with the **Graph Builder**, as shown in Figure 8.10. The **Column Switcher** (available from the toolbar) allows us to easily explore each of the continuous predictors, one after the other.

Figure 8.10: Graph Builder, Validation1 Across mvalue versus lstat

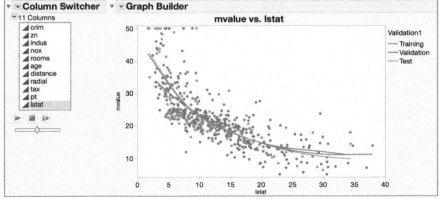

We see that the general relationship between **mvalue** and **lstat** is similar across all three of the validation subsets (Figure 8.10).

We use the same approach for examining the relationship between **mvalue** and **chas** across the three validation subsets (Figure 8.11). The solid lines connect the average of the response, in each validation subset, for **chas** = **0** and **chas** = **1**. This highlights a potential problem. Notice that the average of **mvalue** for **chas = 0** is similar across all three groups, but that the averages are spread apart when **chas = 1**. This is evidence of an interaction between **chas** and **Validation1**. This interaction, which occurred purely by chance, is partly due to the fact that there are very few observations where **chas=1**. This can also occur by chance when data sets are relatively small (we have 506 observations in all, partitioned into three sets).

Figure 8.11:   Graph Builder, Validation1 Across mvalue versus chas

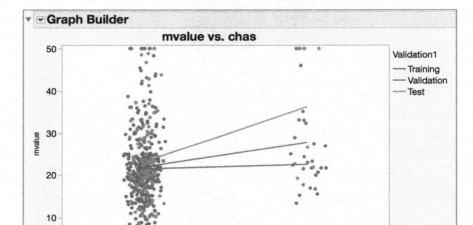

Ideally, we would like the validation role to be independent of the model factors and the response. Because of the potential issue with **Validation1** found above, we create another validation column, **Validation2**, using the same process as before. Examining the distributions of all variables across the validation sets, and the two-way relationships between the factors and the response across the validation sets, we see less evidence of potential issues (see, for example, Figure 8.12). Going forward, we use the **Validation2** column as the validation set. (Note that we ask you to conduct a more thorough exploration of the relationships between the validation sets, the response variable, and the predictor variables in an exercise).

Figure 8.12: **Graph Builder, Validation1 Across mvalue versus chas**

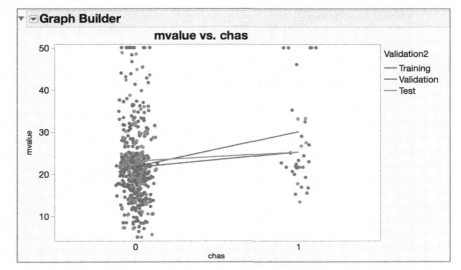

Using Cross-Validation to Build a Linear Regression Model

We now use the validation column (**Validation2**) in the development of models to predict the median home value for the Boston Housing data. The first modeling method that we use is stepwise regression (see Chapter 4 for details on stepwise regression). We specify the model as shown in Figure 8.13, using the **Fit Model** dialog. We enter **mvalue** as the Y variable. In this case, we specify a more complicated model than we will likely need by entering all of the predictor variables as main (linear) effects and also entering all of their two-factor interactions. To do this, we select the predictors from the **Select Columns** list, and select **Factorial to Degree** from the **Macros** menu. The default "degree" that the model specification macro uses is "2", so this adds the twelve factors (or main effects) plus 66 two-way (2nd-order) interaction terms to the list of model effects.

Finally, we specify the **Validation2** column for the **Validation Role**. This helps us determine which terms should be in the final model.

Figure 8.13: Specifying a Stepwise Model with Interactions and Validation

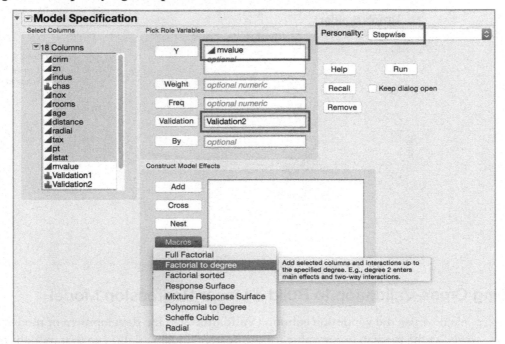

Choosing the Regression Model Terms with Stepwise Regression

We click **Run**, which brings up the stepwise regression dialog (Figure 8.14). Since we are using cross-validation, the stopping rule that determines which model is selected is **Max Validation RSquare**. When we click **Go**, the forward stepwise regression modeling process begins. Terms are added to the model, one by one. Each term that is added to the model term (from the remaining terms) is the one that has the lowest *p*-value. The process continues until there is no observed improvement in the **Validation RSquare**. Figure 8.15 shows the history of terms added to the model. For each step, we see a number of statistics, including the Validation RSquare. At the 18th step, JMP reports that the best Validation RSquare, across all steps, is 0.8002. This corresponds to the Validation RSquare at step 7.

Figure 8.14: Boston Housing Stepwise Initial Dialog

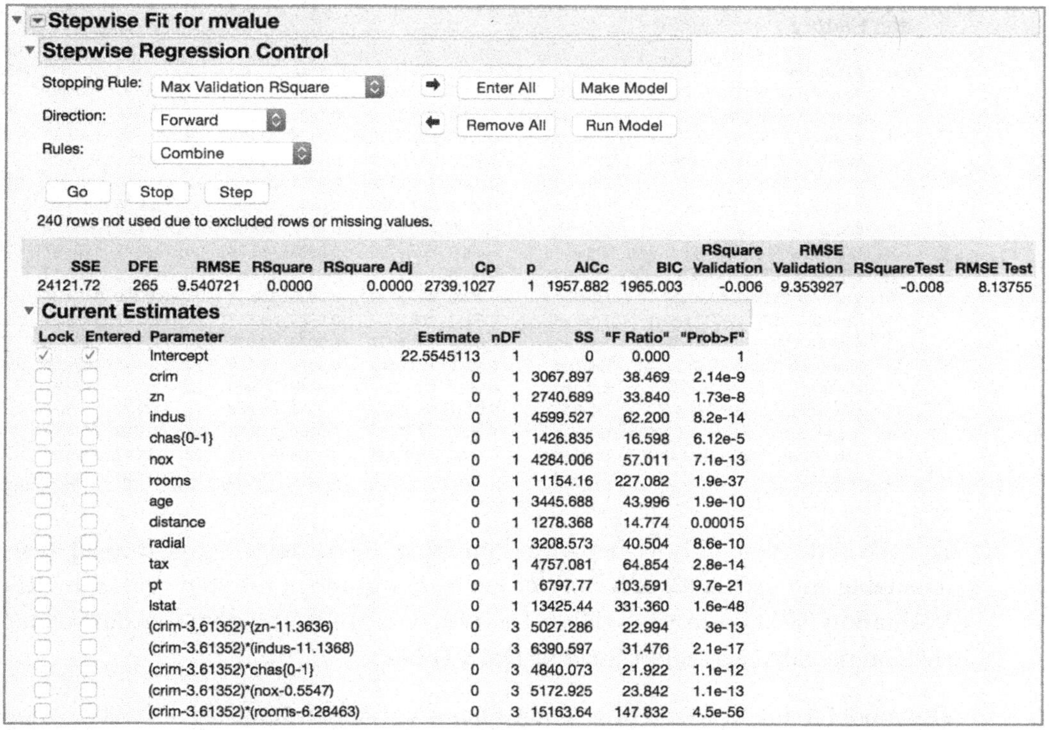

Stepwise Fit for mvalue

Stepwise Regression Control

Stopping Rule:	Max Validation RSquare	⇨	Enter All	Make Model
Direction:	Forward	⬅	Remove All	Run Model
Rules:	Combine			

Go   Stop   Step

240 rows not used due to excluded rows or missing values.

SSE	DFE	RMSE	RSquare	RSquare Adj	Cp	p	AICc	BIC	RSquare Validation	RMSE Validation	RSquareTest	RMSE Test
24121.72	265	9.540721	0.0000	0.0000	2739.1027	1	1957.882	1965.003	-0.006	9.353927	-0.008	8.13755

Current Estimates

Lock	Entered	Parameter	Estimate	nDF	SS	"F Ratio"	"Prob>F"
✓	✓	Intercept	22.5545113	1	0	0.000	1
☐	☐	crim		0	1 3067.897	38.469	2.14e-9
☐	☐	zn		0	1 2740.689	33.840	1.73e-8
☐	☐	indus		0	1 4599.527	62.200	8.2e-14
☐	☐	chas{0-1}		0	1 1426.835	16.598	6.12e-5
☐	☐	nox		0	1 4284.006	57.011	7.1e-13
☐	☐	rooms		0	1 11154.16	227.082	1.9e-37
☐	☐	age		0	1 3445.688	43.996	1.9e-10
☐	☐	distance		0	1 1278.368	14.774	0.00015
☐	☐	radial		0	1 3208.573	40.504	8.6e-10
☐	☐	tax		0	1 4757.081	64.854	2.8e-14
☐	☐	pt		0	1 6797.77	103.591	9.7e-21
☐	☐	lstat		0	1 13425.44	331.360	1.6e-48
☐	☐	(crim-3.61352)*(zn-11.3636)		0	3 5027.286	22.994	3e-13
☐	☐	(crim-3.61352)*(indus-11.1368)		0	3 6390.597	31.476	2.1e-17
☐	☐	(crim-3.61352)*chas{0-1}		0	3 4840.073	21.922	1.1e-12
☐	☐	(crim-3.61352)*(nox-0.5547)		0	3 5172.925	23.842	1.1e-13
☐	☐	(crim-3.61352)*(rooms-6.28463)		0	3 15163.64	147.832	4.5e-56

Figure 8.15: Boston Housing Stepwise Regression Step History

▼ **Step History**

Step	Parameter	Action	"Sig Prob"	Seq SS	RSquare	Cp	p	AICc	BIC	RSquare Validation
1	(rooms-6.28463)*(lstat-12.6531)	Entered	0.0000	18257.84	0.7569	472.04	4	1587.86	1605.55	0.6742
2	(rooms-6.28463)*(pt-18.4555)	Entered	0.0000	861.4629	0.7926	368.79	6	1549.8	1574.45	0.6848
3	(distance-3.79504)*(lstat-12.6531)	Entered	0.0000	602.6045	0.8176	297.77	8	1519.93	1551.47	0.7109
4	(nox-0.5547)*(rooms-6.28463)	Entered	0.0000	341.0061	0.8317	259.31	10	1502.8	1541.18	0.7403
5	(crim-3.61352)*chas{0-1}	Entered	0.0000	521.2242	0.8533	200.42	13	1472.88	1521.37	0.7840
6	chas{0-1}*(tax-408.237)	Entered	0.0002	227.8395	0.8628	176.06	15	1459.68	1514.83	0.7718
7	(radial-9.54941)*(lstat-12.6531)	Entered	0.0002	220.0105	0.8719	152.66	17	1445.97	1507.7	0.8002
8	(rooms-6.28463)*(radial-9.54941)	Entered	0.0000	203.4196	0.8803	129.34	18	1430.17	1495.17	0.7839
9	(age-68.5749)*(distance-3.79504)	Entered	0.0003	184.9772	0.8880	110.31	20	1417.25	1488.72	0.7979
10	(crim-3.61352)*(pt-18.4555)	Entered	0.0029	96.17724	0.8920	100.34	21	1409.99	1484.66	0.7769
11	(crim-3.61352)*(lstat-12.6531)	Entered	0.0013	108.0885	0.8965	88.879	22	1401.11	1478.97	0.7698
12	(age-68.5749)*(radial-9.54941)	Entered	0.0002	138.8827	0.9022	73.589	23	1388.31	1469.33	0.7635
13	(nox-0.5547)*(age-68.5749)	Entered	0.0002	129.0342	0.9076	59.524	24	1375.78	1459.95	0.7590
14	chas{0-1}*(rooms-6.28463)	Entered	0.0017	89.20242	0.9113	50.419	25	1367.37	1454.67	0.7500
15	(crim-3.61352)*(nox-0.5547)	Entered	0.0117	56.08521	0.9136	45.436	26	1362.78	1453.19	0.7513
16	(crim-3.61352)*(rooms-6.28463)	Entered	0.0038	72.10261	0.9166	38.46	27	1355.92	1449.4	0.7746
17	(rooms-6.28463)*(age-68.5749)	Entered	0.0163	48.25812	0.9186	34.452	28	1351.98	1448.53	0.7554
18	Best	Specific	.	.	0.8719	152.66	17	1445.97	1507.7	0.8002

To get a better feel for how stepwise is building the model, we save the step history to a data table and use the **Graph Builder** to create a graph of the **step** versus the **RSquare Validation** (see Figure 8.16). (Hint: To save a table of JMP output to a data table, right-click on the table and select **Save to Data Table**.)

The model that was found after 7 steps has a Validation RSquare of 0.8002. After 10 additional forward steps, the Validation RSquare does not get any larger. So, the model that was found at step 7 is chosen as the best model.

Figure 8.16: Graph of Stepwise Step History

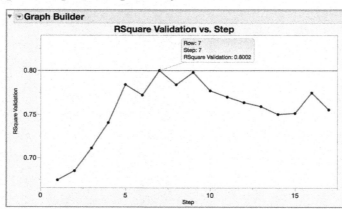

To fit the chosen model, we select **Make Model** in the Stepwise Fit window. This opens the **Fit Model** dialog with the chosen model specified. We then click **Run** to fit the model. The fitted model results for our example are shown in Figure 8.17.

Figure 8.17:   Fitted Model from Stepwise Regression

▼ Response mvalue

▶ **Effect Summary**

▶ **Actual by Predicted Plot**

▼ **Summary of Fit**

RSquare	0.871911
RSquare Adj	0.86368
Root Mean Square Error	3.522577
Mean of Response	22.55451
Observations (or Sum Wgts)	266

▶ **Analysis of Variance**

▼ **Parameter Estimates**

Term	Estimate	Std Error	t Ratio	Prob>|t|
Intercept	44.621362	5.756732	7.75	<.0001*
crim	3.570038	0.716813	4.98	<.0001*
chas[0]	-7.558841	1.263498	-5.98	<.0001*
nox	-13.42288	4.244065	-3.16	0.0018*
rooms	3.2955199	0.450062	7.32	<.0001*
distance	-0.868399	0.193128	-4.50	<.0001*
radial	0.2779489	0.071045	3.91	0.0001*
tax	-0.049922	0.009974	-5.01	<.0001*
pt	-0.690317	0.132594	-5.21	<.0001*
lstat	-0.637494	0.058856	-10.83	<.0001*
(crim-3.61352)*chas[0]	-3.69331	0.715882	-5.16	<.0001*
chas[0]*(tax-408.237)	0.0390813	0.009534	4.10	<.0001*
(nox-0.5547)*(rooms-6.28463)	-11.5764	4.234603	-2.73	0.0067*
(rooms-6.28463)*(pt-18.4555)	-0.881481	0.161285	-5.47	<.0001*
(rooms-6.28463)*(lstat-12.6531)	-0.33509	0.049402	-6.78	<.0001*
(distance-3.79504)*(lstat-12.6531)	0.0405233	0.023826	1.70	0.0902
(radial-9.54941)*(lstat-12.6531)	-0.014773	0.004937	-2.99	0.0030*

▶ **Effect Tests**

▼ **Crossvalidation**

Source	RSquare	RASE	Freq
Training Set	0.8719	3.4082	266
Validation Set	0.8002	4.1688	127
Test Set	0.7195	4.2928	113

Making Predictions

We save the model results. This creates a new column in the table, **Pred Formula mvalue** (use the red triangle and then select **Save Columns > Prediction Formula**). This saved column contains the formula that calculates the predicted value for each row in the data table, and we can use this formula to predict new values. As we will see, we can also use this saved prediction formula to compare the performance of this model to other models that we have created.

Using Cross-Validation to Build a Decision Tree Model

Even though the process described above helped us choose the "best" linear regression model, the final chosen model is just one of many options available to build a prediction model. Decision trees can sometimes have better predictive properties than regression models, so we fit a simple decision tree model to predict **mvalue**. Again, we use cross-validation to guide the decision tree building process.

In this situation, the splits in the decision tree are determined by the LogWorth statistic, as is described in Chapter 6, "Decision Trees." Here, we again use **Validation2** to specify the training, validation, and test subsets, rather than using the **Validation Portion** (as we did in Chapter 6). After each split of the decision tree, the validation RSquare is calculated. When the validation RSquare stops improving, the decision tree automatically stops splitting after 10 additional splits.

Figure 8.18 shows the **Partition** platform launch dialog, where we choose **mvalue** as the response and all of the potential predictors as the factors. **Validation2** is chosen as the validation role column.

Figure 8.18:   Boston Housing Partition Dialog with Validation2

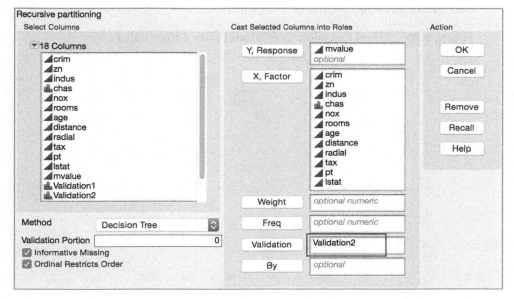

The initial model fit report is shown in Figure 8.19. Since we are using model cross-validation, JMP reports Training, Validation, and Test RSquare. We click **Go** to begin the automatic decision tree building process.

Figure 8.19:   Boston Housing Initial Partition Output with Cross-Validation

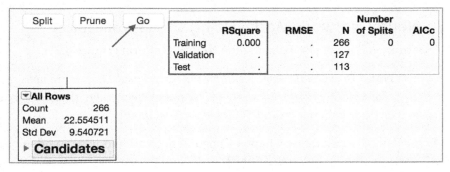

In Figure 8.20, we see the statistics for the final model and the **Split History**. The bottom line in the **Split History** graph corresponds to the validation RSquare, the middle is the test RSquare, and the top line is the training RSquare. The largest tree that was built had 29 splits. However, JMP stops adding branches to the tree after ten splits steps if there is no further improvement in validation RSquare. The chosen decision tree model, with the best validation RSquare, has only 19 splits.

Figure 8.20: Fitted Decision Tree Model with Cross-Validation

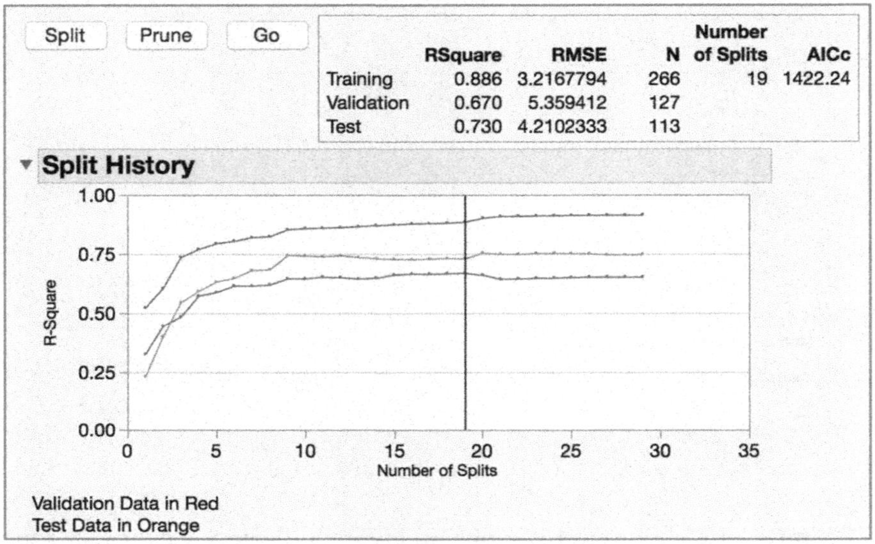
Split
Prune
Go
RSquare RMSE N Number of Splits AICc
Training 0.886 3.2167794 266 19 1422.24
Validation 0.670 5.359412 127
Test 0.730 4.2102333 113
Split History
R-Square
1.00
0.75
0.50
0.25
0.00
0 5 10 15 20 25 30 35
Number of Splits
Validation Data in Red
Test Data in Orange

We use the chosen decision tree model and save the prediction formula to the data table (use the top red triangle, **Save Columns > Save Prediction Formula**). This creates a new column, **mvalue Predictor.**

Fitting a Neural Network Model Using Cross-Validation

A third model is developed using the **Neural** platform (Figure 8.21).

> ***Note:*** *To obtain the same result for the Neural model in this example, set the random seed to 1000 prior to launching the Neural platform using the Random Seed Reset add-in or the JMP Script Neural(Set Random Seed(1000));.*

As we have discussed (see Chapter 7) neural network models can often outperform other modeling methods due to their flexibility, but they are also prone to over-fitting. Because of this, some form of cross-validation is required when building neural network models in JMP.

Figure 8.21:   Boston Housing Neural Model Launch Dialog with Cross-Validation

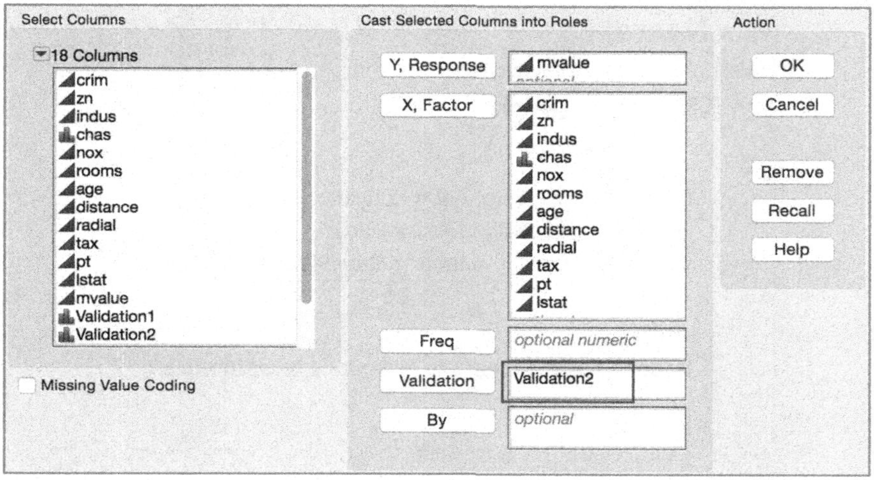

We run the default single-layer model with three "sigmoid" TanH nodes (see Figure 8.22). In the model fitting process, cross-validation is used to determine the best value for the tuning parameter that is part of the neural model fitting process. The purpose of this tuning parameter is to help prevent over-fitting and provide a better performing model.

Figure 8.22:   Boston Housing Neural Model Launch

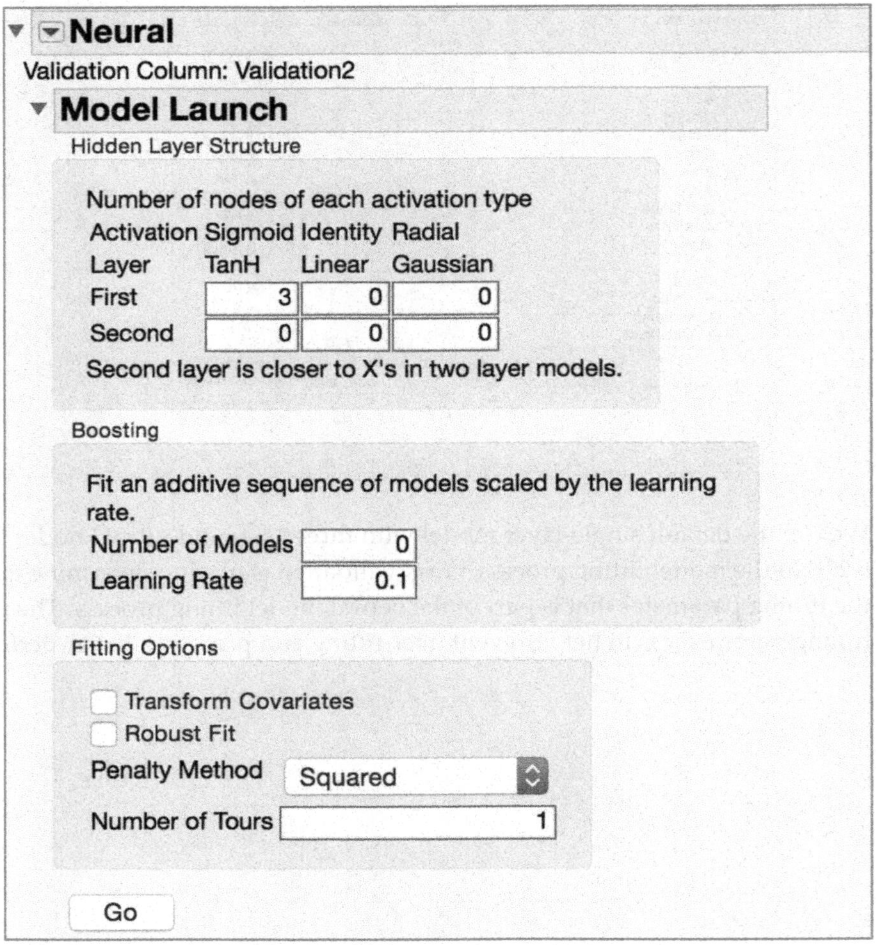

▼ ⊟ **Neural**
Validation Column: Validation2
▼ **Model Launch**

Hidden Layer Structure

Number of nodes of each activation type
Activation Sigmoid Identity Radial

Layer	TanH	Linear	Gaussian
First	3	0	0
Second	0	0	0

Second layer is closer to X's in two layer models.

Boosting

Fit an additive sequence of models scaled by the learning rate.

Number of Models	0
Learning Rate	0.1

Fitting Options

☐ Transform Covariates
☐ Robust Fit

Penalty Method	Squared ⌃⌄
Number of Tours	1

Go

The results for the fitted model, along with the model diagram, are shown in Figure 8.23. There are a number of model-fitting options available within the **Neural Model Launch** dialog. We discuss these options in Chapter 9.

As with the previous two models (stepwise regression and decision tree), we save the prediction formula to the data table (from the fitted model's red triangle, choose **Save Columns > Save Profile Formulas**). The new column is named **Predicted mvalue**.

Figure 8.23:   Boston Housing Neural Fitted Model with Cross-Validation

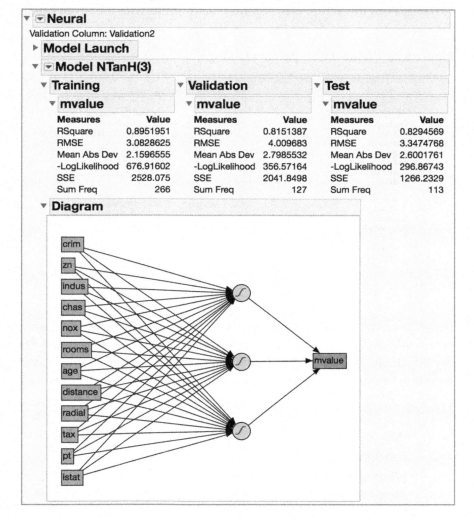

Model Comparison

We have fit three different models using cross-validation, and have saved the prediction formulas for the three models to the data table. In JMP Pro, we use the **Model Comparison** platform to compare the performance of these models (select **Analyze > Modeling > Model Comparison**). We specify the columns with the prediction formulas for each model as **Y, Predictors**, and select **Validation2** as the **Group** variable. (Note that an alternative is to leave the **Y, Predictors** field blank, and use **Validation2** as the **By** variable. With this setup, JMP will find the columns with the saved prediction formulas for you.)

Figure 8.24:   Model Comparison Dialog

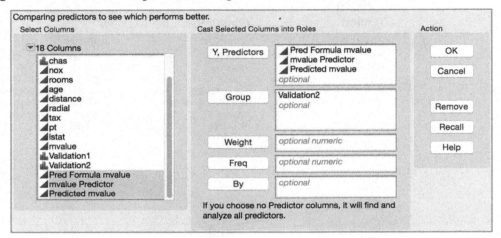

In the resulting Model Comparison report (Figure 8.25), we use the statistics from the test set for model comparison. The Neural Network model has the highest RSquare value (0.8295). It also has the lowest Root Average Squared Error (RASE). But the Average Absolute Error (AAE) is slightly higher than the AAE for the Least Squares model.

Figure 8.25:   Model Comparison Report

▼ ☑ Model Comparison
▶ Predictors
▼ Measures of Fit for mvalue

Validation2	Predictor	Creator	.2 .4 .6 .8	RSquare	RASE	AAE	Freq
Training	Pred Formula mvalue	Fit Least Squares		0.8719	3.4082	2.4207	266
Training	mvalue Predictor	Partition		0.8859	3.2168	2.1457	266
Training	Predicted mvalue	Neural		0.8952	3.0829	2.1597	266
Validation	Pred Formula mvalue	Fit Least Squares		0.8002	4.1688	2.8230	127
Validation	mvalue Predictor	Partition		0.6697	5.3594	3.3003	127
Validation	Predicted mvalue	Neural		0.8151	4.0097	2.7986	127
Test	Pred Formula mvalue	Fit Least Squares		0.7195	4.2928	2.5259	113
Test	mvalue Predictor	Partition		0.7302	4.2102	3.2313	113
Test	Predicted mvalue	Neural		0.8295	3.3475	2.6002	113

If we have to choose only one of these models from the perspective of performance and prediction error, we would select the neural network model.

Additional options for comparing the models, such as the **Profiler**, are available from the top red triangle in the model comparison window. For models with categorical responses, the misclassification rate and other relevant statistics are reported, and options such as the **ROC Curve**, the **Lift Curve**, and the **Confusion Matrix** are available from the top red triangle.

Key Take-Aways

The use of cross-validation is central to our ability to develop good predictive models that neither over-fit nor under-fit our data. As such, it is one of the most important concepts in predictive modeling. In many situations, we use two hold-out subsets: training and validation. If we also hold out a test subset, then we have the ability to assess model performance on an unbiased sample that wasn't used in the model building or selection process. This allows for good model evaluation, and makes it easy to compare and choose among competing models.

Exercises

Exercise 8.1:   This exercise will give you practice in creating training, validation, and test subsets, and on evaluating the characteristics of these subsets.

a. Open the example data set **Boston Housing BBM Ch9.jmp**. Use the **Make Validation Column** utility to create random training, validation, and test subsets. Use 50% of the data for training, 25% for validation, and 25% for test.

   Note: To produce the same validation set as others, set the random seed to 1000 before creating the validation column.

b. Evaluate the training, validation, and test subsets, checking to see that the response variable and predictor variables have similar distributions across each subset, and also check for interactions between the predictor variables and subsets.

c. Create a new validation column if needed (use the random seed 1001).

d. Save the file with the validation column. We'll use this in an exercise to follow.

Exercise 8.2: Use **Boston Housing BBM Ch9.jmp** for this exercise.

a. Create final prediction models, with the validation column that you created in Exercise 8.1, using the following modeling methods:

1. Forward Stepwise Regression
2. Decision Tree
3. Neural Network

b. Compare the final models using the **Model Comparison** platform.

1. Using the default statistics provided, is there one model that stands out as being better?
2. What criteria did you use to determine the best performing model?
3. Describe the criteria that you used in part 2.
4. Explore the options available under the red triangle. Explain how the Profiler and first two plots can be useful in comparing and selecting the best performing model.
5. Explain the value of using a test set. Why is it important to use a hold-out sample for model evaluation and selection?

Exercise 8.3: Repeat Exercise 8.1, using the data set **Bank Marketing BBM Ch8.jmp**.

The response is the binary yes/no variable **y**, "has the client subscribed a term deposit?". Determine whether the **Stratified Random** option should be used. Again, set the random seed to 1000 before creating the validation column.

The variables in this data set (Moro and Rita, 2014) and background information are provided at https://archive.ics.uci.edu/ml/datasets/Bank+Marketing.

Exercise 8.4: Use **Bank Marketing BBM Ch8.jmp**, for this exercise.

a. Find prediction models, with the validation column that you created in Exercise 8.3, using the following modeling methods:

1. Forward Stepwise Regression. Note that, because of the size of the data set and the computational complexity of logistic regression, it is probably better to include only main effects (no interaction terms) for the predictor variables.
2. Backward Stepwise Regression
3. Decision Tree
4. Neural Network

b. Compare the final models using the **Model Comparison** platform.

   1. Is there one model that stands out as performing the best on the test set?

   2. What criteria did you use to determine the best performing model?

   3. Describe the criteria that you used in part 2.

   4. Explore the options available under the top red triangle. Explain how the first group of options (the first 5) can be used to compare competing models.

Exercise 8.5: Use the **Credit Card Marketing BBM.jmp** data for this exercise. This data set is described in Case Study 2 in Chapter 6. The response is **Offer Accepted**.

a. Create a distribution of **Offer Accepted**. The target category is **Yes**. What percent of the offers were accepted?

b. Use the create validation column utility to create training, validation, and test sets, stratified on **Offer Accepted**.

c. Explore this validation column as you did in exercises 8.1 and 8.3. What percent of each set is **Offer Accepted = Yes** and **Offer Accepted = No**? Are there any issues with using this validation column?

d. Create a decision tree and a neural model to predict **Offer Accepted** using this validation column, and then compare these models using the **Model Comparison** platform.

e. Which is your best performing model? What is the misclassification rate?

References

Box, George E. P., and Norman R. Draper. 1987. *Empirical Model-Building and Response Surfaces*, p. 424. Wiley.

Harrison, David, and Daniel Rubinfield. 1978. "Hedonic Prices and the Demand for Clean Air", *Journal of Environmental Economics and Management*, 5, 81-102.

Moro, S., P. Cortez, and P. Rita. 2014. *A Data-Driven Approach to Predict the Success of Bank Telemarketing. Decision Support Systems*. Elsevier, 62:22-31.

Advanced Methods

Overview

Our approach in Chapters 3 through 8 has been to provide a fundamental understanding of several core modeling techniques, and to illustrate how to apply these methods to solve business problems. For many of these techniques, there are more advanced model-building methods available, and using these advanced methods can result in improved predictive ability. In this chapter, we provide an overview of several of these more advanced methods.

We use the word "advanced" in this chapter to mean several things:

- Models that have much more complexity and that provide better predictive ability.

- Regression models when there are a large number of possible predictors, but only a few that are likely to really be important.

- Model fitting techniques that are more modern and that require computationally advanced fitting algorithms.

- Models that are the averages or sums of many other models.

This chapter covers the following advanced methods that are available in JMP Pro:

- Partition Methods (Random Forest and Boosted Tree)

- Boosted Neural Networks

- Generalized Regression (Forward Stepwise Regression, Lasso, Elastic Net, and Ridge Regression)

Concepts in Advanced Modeling

Three important concepts that are used in these advanced models are *bagging*, *boosting*, and *regularization*.

> **Note:** *What follows is a little more mathematically complex than what we've covered in the text thus far. JMP Pro handles all the difficult calculations, but we provide a basic overview of the underlying algorithms to help you understand each of the modeling methods.*

Bagging

Bagging is a shorthand term for a technique known as *bootstrap aggregation*. Bagging is a model averaging method where several models of the same type are built on *bootstrap* samples drawn from the data. A bootstrap sample is a sample of the same size as the original data, but drawn with replacement. The term *sampling with replacement* means that each individual observation may not be sampled at all, or that an individual observation could be sampled one or more times.

The general algorithm for bagging is:

1. Repeat the following *B* times:
    a. From the data table, with the number of training rows *N*, select a random sample with replacement from the training data, also of size *N*. This is a *bootstrap* sample.
    b. Fit a model on the bootstrap training data, and save the model prediction equation.
2. Average all *B* of the models generated from the bootstrap samples to create the bootstrap aggregated (bagged) model.

Bagging is particularly useful for models that are, on average, unbiased, but that tend to have high variability associated with their predictions. Using a model that is the average of several unbiased models built on bootstrap samples leads to a model that is still unbiased, but that is expected to have a lower amount of error variability. Bagging is applied specifically in the **Bootstrap Forest** technique in JMP Pro, which we describe later in this chapter.

Boosting

Boosting is a model building method that is an additive modeling approach. The basic idea behind boosting is that if we fit a simple model and make predictions with this model, we might be able to explain the remaining unpredicted variation with another model. If we build a sequence of models, with each model successively predicting more of the remaining residual error, we can eventually arrive at a good overall model.

Boosting builds a sequence of several low-complexity, poorly predicting models that are added together to arrive at the final model. The individual models that are summed together are called "layers" of the boosted model, and each layer attempts to predict a small portion of the remaining residual error.

While boosting is conceptually simple, the mathematics and algorithms behind boosting can be complex. We will not go into the particular boosting algorithms used in JMP Pro with much depth.

Boosting is used in the **Partition Platform** to create **Boosted Tree** models, and in the **Neural Platform** to create **Boosted Neural** network models.

Regularization

Regularization is an approach to model fitting, where instead of just using the traditional approach of minimizing the SSE or the -LogLikelihood, you include a penalty term that is a function of the parameter estimates. For instance, when the response is continuous, the model fitting process finds the best fitting model that minimizes the -LogLikelihood plus a penalty that is a function of the parameters in the model. If we let the parameters in the model be denoted by $(b_1, b_2, \ldots, b_k)$, the estimates of the model parameters are chosen so that we minimize

$$G(b_1, b_2, \ldots, b_k) = L(b_1, b_2, \ldots, b_k) + \lambda R(b_1, b_2, \ldots, b_k)$$

where:

- $G(b_1, b_2, \ldots, b_k)$ represents the overall generalized "loss" function that is minimized to find the best parameter estimates,

- $L(b_1, b_2, \ldots, b_k)$ is the "negative log-likelihood" that is generally related to the prediction error,

- λ is the penalty (tuning) parameter that controls how much overall penalty is applied, and

- $R(b_1, b_2, \ldots, b_k)$ is the regularization penalty function.

The type of penalty applied depends upon the modeling method chosen. The tuning parameter λ is chosen by picking the value that gives the best model predictions.

The penalty $R(b_1, b_2, \ldots, b_k)$ is generally related to the number of parameters and the overall magnitude of the parameters. The more parameters and the larger the values of the parameters, the greater the penalty that is applied. What the penalty does, in effect, is force some or all of the parameters in the model to be closer to zero than they would be if chosen without using a penalty. As a result, this approach is also known as *shrinkage estimation* (it tends to shrink the parameter estimates closer to zero).   This shrinking of the parameters in the model can lead to very efficient model selection. This approach to model selection can perform better than stepwise regression (introduced in Chapters 4 and 5) in terms of overall predictive ability.

There are several types of regularization penalties that can be applied, which lead to different regularized (or *penalized*) modeling methods. Each of the following regularized methods is available from the **Generalized Regression** personality in the **Fit Model** platform in JMP Pro.

1. *Ridge Regression*: The penalty is based on the sum of the squared parameter values, $R(b_1, b_2, \ldots, b_k) = \sum_i^k b_i^2$.

2. *LASSO*: The penalty is based on the sum of the absolute parameter values, $R(b_1, b_2, \ldots, b_k) = \sum_i^k |b_i|$.

3. *Elastic Net*: uses a weighted combination of Ridge (sum of squared parameter values) and LASSO (sum of absolute parameter values), with an additional tuning parameter (α), $R(b_1, b_2, \ldots, b_k) = (1-\alpha)\sum_i^k b_i^2 + \alpha \sum_i^k |b_i|$.

Note: The Neural platform also utilizes parameter regularization, similar to either ridge regression or LASSO. This was discussed in Chapter 7.

Advanced Partition Methods

Building decision tree models with the **Partition** platform was covered in Chapter 6. Decision trees can perform very well for predictive modeling and for uncovering relationships between factors and a response, but they do have some drawbacks:

1. Even though decision trees tend to have low bias (on average, they are expected to predict the target response), they tend to have higher prediction variance (the expected prediction error).

2. Decision trees can become "dominated" by the factors that have the biggest correlation with the response, and other potentially useful factors may be underrepresented or completely ignored when building the model.

3. Decision trees can become large and computationally slow in computing predicted values.

In the JMP Pro **Partition** platform, two other types of decision tree models that help overcome these drawbacks are available: **Bootstrap Forest** and **Boosted Tree**.

Bootstrap Forest

The bootstrap forest model is built using bagging and random sampling of the factors to build a predictive model. The general algorithm for the bootstrap forest method is:

1. Repeat *B* times:

   a. Draw a bootstrap sample from the data set.

   b. Build a large decision tree model on the bootstrap sample, but with one modification: At every split decision, instead of considering all of the factors to find the optimum split, consider only a random subset of the factors.

   c. Save the decision tree model, T_b.

2. The bootstrap forest prediction model is just the average of all *B* of the tree models. The overall predictive model is: $BF = \dfrac{1}{B}\sum_{b=1}^{B} T_b$.

By averaging over all of the decision trees chosen in this way, you can arrive at an aggregated model that can predict much better that any single decision tree model.

Bootstrap Forest Example

To illustrate how to build a bootstrap forest model, we use the **Titanic Passenger BBM Ch9.jmp** data (the Titanic Passenger example was introduced in Chapter 5).

We populate the **Partition** dialog as we have seen, and choose **Bootstrap Forest** as the method (top, in Figure 9.1).

> *Note: To recreate the Bootstrap Forest output shown here, use the Random Seed Reset add-in to set the seed value to 2000.*

In the resulting **Bootstrap Forest Specification** window (bottom, in Figure 9.1), we specify the bootstrap model fitting values.

The bootstrap forest fitting options are described in the next section.

Figure 9.1:   Bootstrap Forest Partition Dialog and Specification

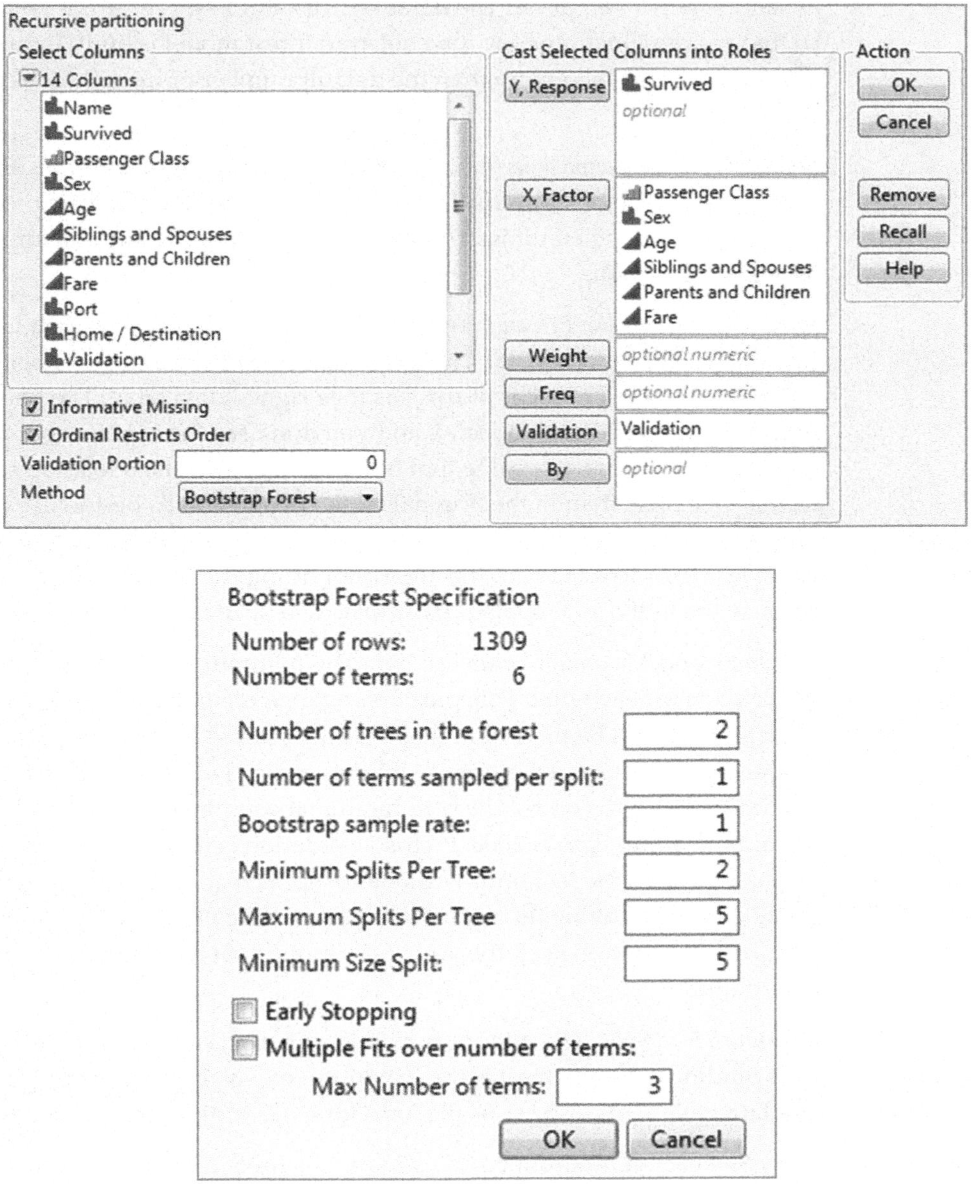

Bootstrap Forest Fitting Options

1. *Number of trees in the forest*: The default is 100, but in this example, we use just 2 so that we can illustrate how the bootstrap forest model is built. In practice, you may want to use even more than the default number of trees (hundreds or even thousands).

2. *Number of terms sampled per split*: For each split decision in each tree, this is the number of predictor columns that are randomly selected to participate in the split decision. The default is ¼ of the total number of predictor columns specified in the launch dialog.

3. *Bootstrap sample rate*: For each tree, this is the proportion of the data table that is sampled (with replacement). The default is 1, which means the bootstrap sample will have the same number of rows as the original data table. The bootstrap sampling happens automatically, and you don't see the separate bootstrap samples. If you choose a value that is less than 1, then the bootstrap sample will have fewer rows than in the original table. In general, it's best to use the default here. In some situations, such as when you have a very large data set, using a value less than 1 can lead to the algorithm fitting the model more quickly because the individual bootstrap samples will be of smaller size.

4. *Minimum* and *Maximum Splits Per Tree*: The minimum and maximum number of splits made for each tree. This puts a size range on each random tree in the bootstrap forest. The minimum helps ensure at least a moderate amount of model complexity, and the default minimum size is 10 splits per tree. The maximum helps to control the computational complexity of the model, and the default maximum size is 2000. If cross-validation is used, an individual tree may stop earlier than the maximum number of splits if adding more splits to the tree is no longer improving its validation RSquare. The next setting, *Minimum Split Size*, will also tend to keep the size of the individual trees smaller than the maximum.

5. *Minimum Size Split*: The minimum number of observations needed to qualify as a split candidate. The default is the maximum of 5 or 1% of the number of rows in the data table. This also keeps the individual trees from becoming too over-fit.

6. *Early Stopping*: If hold-out cross-validation is used, then a check box is displayed and selected by default. Early stopping will cause JMP Pro to stop adding trees to the bootstrap forest if the validation performance of the model stops improving.

7. *Multiple Fits over number of terms*: If selected, JMP will fit multiple bootstrap forests, with each bootstrap forest using a progressively larger value for *Number of terms sampled per split*.

For this example, we have selected a small number of trees and splits so that the structure of the model can be seen and understood easily. The bootstrap forest model is composed, in this case, of two trees, shown in Figure 9.2. Options for viewing trees are available under the top red triangle, **Show Trees**. Here, we select **Show names categories** to show a small tree view (we've moved the trees side-by-side for comparison).

Figure 9.2: Bootstrap Forest Trees

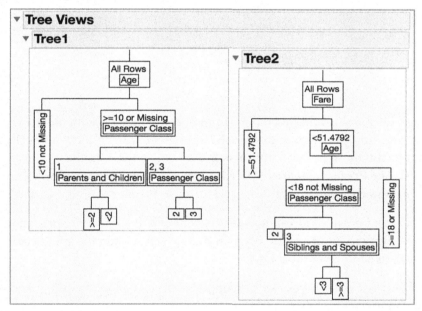

The prediction formula for **Prob(Survive==Yes)** is the average of the prediction formulas for the two trees, as shown in Figure 9.3.

Figure 9.3:   Bootstrap Forest Prediction Formula

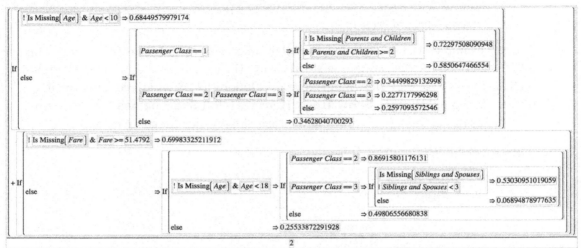

In practice, we would fit a much larger bootstrap forest, with more trees and more splits per tree. Interpretation of this much larger model can be challenging, but the **Column Contributions** report can give insight into which terms in the model are contributing the most. The **Profiler** can also be used to interact with the prediction formula and explore model effects (save the formula, and then use **Profiler** from the **Graph** menu).

Note that bootstrap forest models will generally include more predictors than decision trees. Since terms are randomly sampled for selection in the split decision, more terms are allowed to enter the model.

In Exercises 1 and 2, you have an opportunity to explore the fitting and interpretation of bootstrap forest models.

Boosted Tree

JMP Pro can apply boosting to decision tree models, and the resulting model that is generated is called a *boosted tree*. A boosted tree consists of decision trees that are built successively, in layers. These layers consist of decision trees that are typically very small, often with only a few splits in each tree. The boosting algorithm can add hundreds of these smaller trees together, creating an additive or *"ensemble"* model that can be a very good predictive model. The overall complexity of boosted trees tends to be much lower than bootstrap forest trees, but the algorithm can often result in better predictive ability.

For problems with very large data sets, boosted trees can be fit to the data much more quickly, and the computational load for making model predictions can be much lower.

Boosted Tree Example

We again use the **Titanic Passenger BBM Ch9.jmp** data set, and illustrate fitting a boosted tree model. Figure 9.4 shows the **Partition** platform for launching a boosted tree model (top) and the intermediate boosted tree model specification dialog.

The boosted tree fitting options are described in the next section.

Figure 9.4:   Boosted Tree Model Dialog and Specification

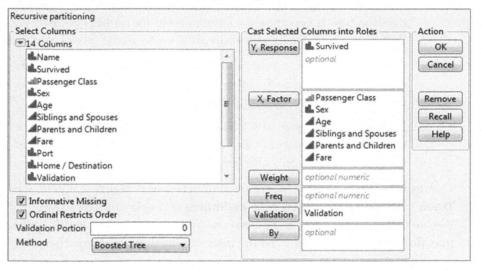

Boosted Tree Fitting Options

1. *Number of Layers*: The maximum number of layers to include in the final tree. The default value is 50, but for this simple example we use only 3 layers.

2. *Splits Per Tree*: The number of splits for each layer. The default value is 3.

3. *Learning Rate*: A value between 0 and 1. This value controls how much each layer contributes to the overall model, and the default value is 0.1.

4. *Overfit Penalty*: (only if categorical) A parameter that helps protect against fitting probabilities equal to zero. The default value is usually best to use here.

5. *Minimum Size Split*: The smallest number of rows that can be in any branch of the decision tree. If you choose a large number for number of *Splits Per Tree*, this setting can help prevent the individual layers in the model from being too large.

6. *Early Stopping*: This option, which is available and checked by default if cross-validation is used, will cause JMP Pro to stop adding layers earlier if the validation RSquare stops improving.

7. *Multiple Fits over splits* and *learning rate*: This creates a boosted tree for several combinations of *Splits Per Tree* and *Learning Rate*. Minimum values are those specified above in *Splits Per Tree* and *Learning Rate*.

The resulting three boosted layers are shown below in Figure 9.5 (here, we use **Show Trees > Show names categories estimates** to display the trees with model estimates). We see that predictors **Sex** and **Fare** show up in each tree, but in different ways in each tree. (In Figure 9.5, we've moved the trees side-by-side for comparison.)

Figure 9.5: Boosted Tree Views

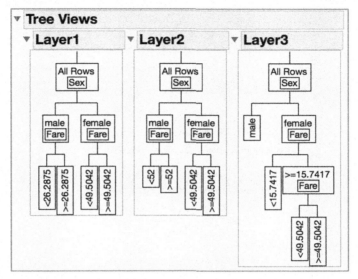

Remember that for models that predict categorical outcomes, the resulting models predict the probability of occurrence for each outcome based on the values of the predictors. If you look closely at the trees in Figure 9.5, you may notice something odd. The "Estimate" value in each terminal node is not a probability, but a general estimate. What is happening here is something that is particular to the boosting algorithm when used to predict a categorical response: The layers in the boosted tree estimate the *logit* of the probability (refer to Chapter 5 for details on the definition and interpretation of the logit).

The prediction formula (Figure 9.6) for this model has the logistic transformation applied to the *sum* of the boosted decision tree layers.

Figure 9.6:   Boosted Tree Prediction Formula for Prob[Survived=No]

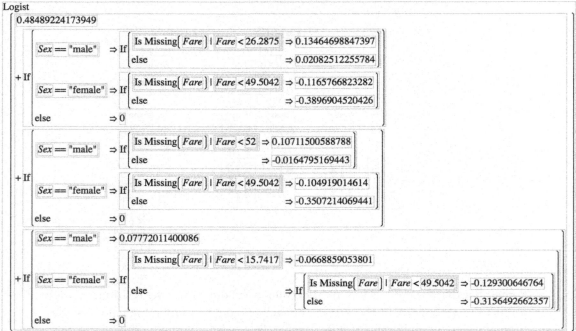

Boosted Neural Network Models

We introduced neural networks in Chapter 7. One option that we did not explore, however, was applying boosting to a neural model.

We return to the first example used in Chapter 7: the Churn data (file **Churn BBM Ch9.jmp**). We populate the **Neural** model dialog as shown in Figure 9.7.

Figure 9.7: Neural Network Dialog

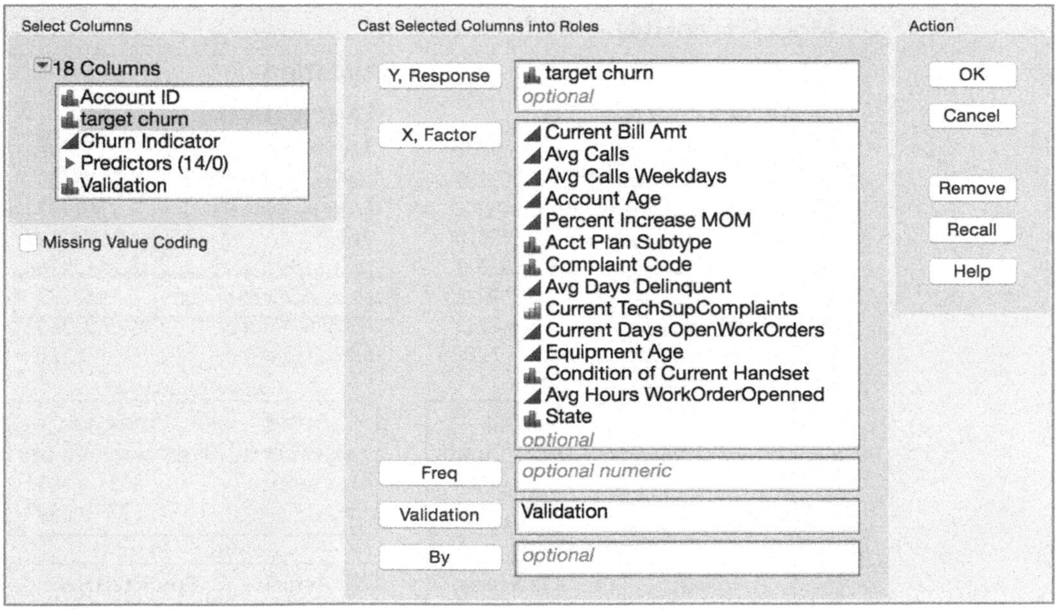

As a reminder, the **Model Launch** dialog provides options for the number and type of nodes for the two layers in the model. For illustration, we begin with a simple single-layer model with one TanH node (Figure 9.8).

The model results are shown in Figure 9.9, and the neural network diagram for the model is shown in Figure 9.10.

For this model design, we have about a 19.6% validation misclassification rate.

Figure 9.8: Neural Network Specification

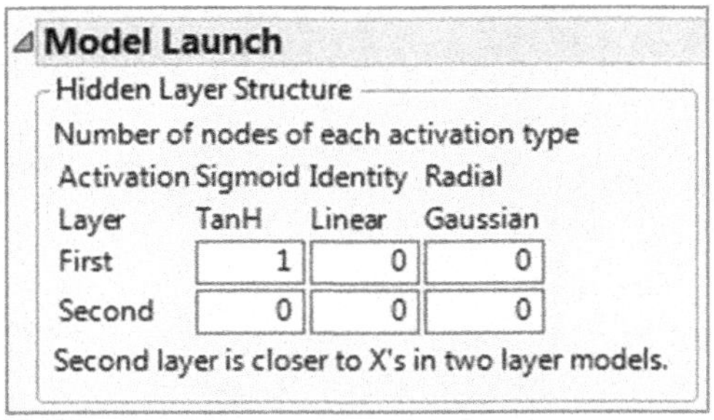

Figure 9.9:   Neural Network Results

Model NTanH(1)

Training

target churn

Measures	Value
Generalized RSquare	0.5109389
Entropy RSquare	0.348709
RMSE	0.3767526
Mean Abs Dev	0.2843671
Misclassification Rate	0.1909028
-LogLikelihood	1308.7713
Sum Freq	2902

Confusion Matrix

Actual	Predicted	
target churn	No Churn	Churn
No Churn	1261	244
Churn	310	1087

Confusion Rates

Actual	Predicted Rate	
target churn	No Churn	Churn
No Churn	0.838	0.162
Churn	0.222	0.778

Validation

target churn

Measures	Value
Generalized RSquare	0.5123538
Entropy RSquare	0.3501298
RMSE	0.376819
Mean Abs Dev	0.2862156
Misclassification Rate	0.1957237
-LogLikelihood	546.51785
Sum Freq	1216

Confusion Matrix

Actual	Predicted	
target churn	No Churn	Churn
No Churn	531	111
Churn	127	447

Confusion Rates

Actual	Predicted Rate	
target churn	No Churn	Churn
No Churn	0.827	0.173
Churn	0.221	0.779

Figure 9.10: Neural Network Diagram

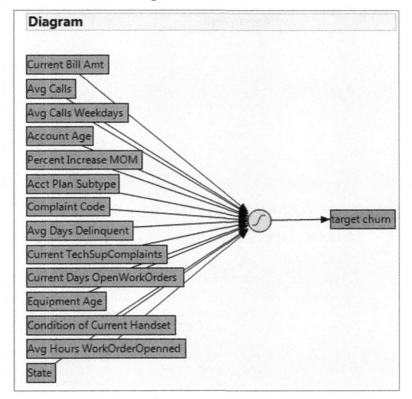

Now we apply boosting to this model. We specify that we would like up to 500 layers in the boosted neural model (**Number of Models**), and we use the default **Learning Rate** of 0.1.

> **Note**: *The default number of layers is 50, but we have chosen 500 to allow the algorithm to go further than 50 layers if needed*

The results are shown in Figure 9.12. JMP Pro stops at 26 layers, since adding additional layers doesn't improve the boosted model's validation RSquare.

Figure 9.11:   Neural Network Third Model Specification

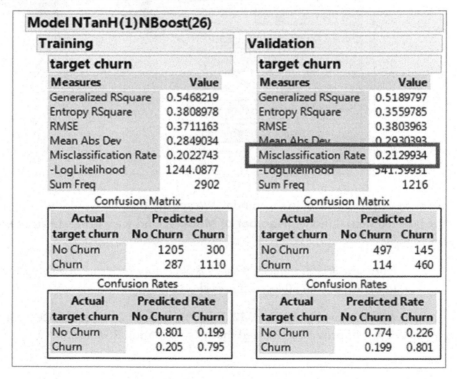

Figure 9.12:   Boosted Neural Network Results

Model NTanH(1)NBoost(26)

Training		Validation	
target churn		**target churn**	
Measures	**Value**	**Measures**	**Value**
Generalized RSquare	0.5468219	Generalized RSquare	0.5189797
Entropy RSquare	0.3808978	Entropy RSquare	0.3559785
RMSE	0.3711163	RMSE	0.3803963
Mean Abs Dev	0.2849034	Mean Abs Dev	0.2930393
Misclassification Rate	0.2022743	Misclassification Rate	0.2129934
-LogLikelihood	1244.0877	-LogLikelihood	541.59931
Sum Freq	2902	Sum Freq	1216

Confusion Matrix			Confusion Matrix		
Actual	**Predicted**		**Actual**	**Predicted**	
target churn	**No Churn**	**Churn**	**target churn**	**No Churn**	**Churn**
No Churn	1205	300	No Churn	497	145
Churn	287	1110	Churn	114	460

Confusion Rates			Confusion Rates		
Actual	**Predicted Rate**		**Actual**	**Predicted Rate**	
target churn	**No Churn**	**Churn**	**target churn**	**No Churn**	**Churn**
No Churn	0.801	0.199	No Churn	0.774	0.226
Churn	0.205	0.795	Churn	0.199	0.801

The validation misclassification rate is slightly higher (21.3% versus 19.6%) than what we saw with the initial neural model. However, since the threshold for classifications is 0.5, the misclassification rate probably isn't the best measure of model performance. We refer the reader to the discussion of *ROC* and *Lift* curves in Chapter 6.

If we examine the network diagram for this model, we see that it appears to be like a neural model with 26 nodes in the first hidden layer. This is because we can, with some mathematical rearrangement, represent the boosted layers in the models as if they were separate nodes in the hidden layer. In essence, the boosting procedure has grown the neural network automatically for us.

Figure 9.13: Boosted Neural Network Model Diagram

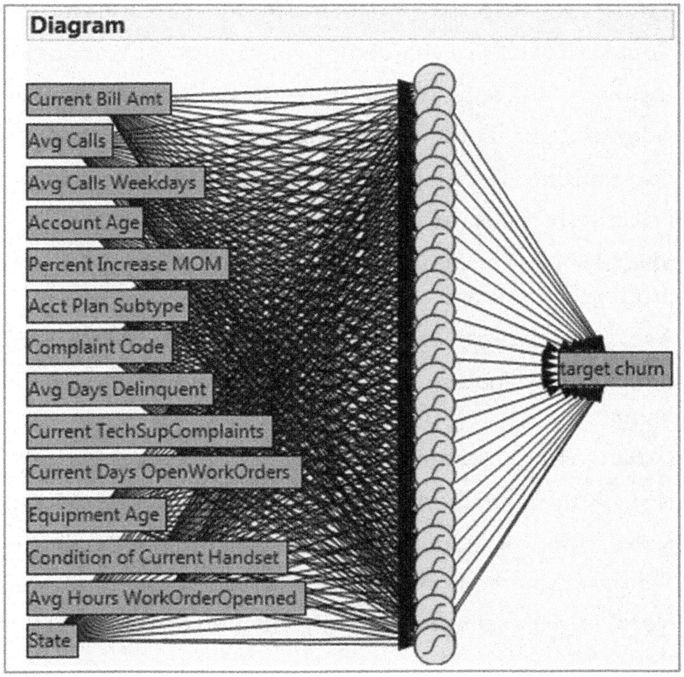

Generalized Regression Models

The **Generalized Regression** platform in JMP Pro encompasses several types of regression models. These models can be used to predict either continuous or categorical response variables. You can also choose among several residual error distributions that fit the expected variation that you see in your data. This provides a large variety of models that you can fit to match the problem that you are trying to solve. In addition, **Generalized Regression** incorporates the regularized (penalized) regression model-fitting methods introduced earlier in this chapter: Ridge regression, Lasso, and Elastic Net.

In this section we provide examples of using the **Generalized Regression** platform to fit these models.

The Data   Car Seats BBM Ch9.jmp

The data set contains simulated data on sales of a brand of car seat. There are 400 rows of data, corresponding to 400 stores where this product was sold.

> **Sales:**   The response variable, unit sales (in thousands)
>
> **CompPrice:**   Price charged by competitor for comparable brand
>
> **Income:**   Average income of the community the store is located in
>
> **Advertising:**   Local advertising spend
>
> **Population:**   Population in the region the store is located in
>
> **Price:**   The sales price for the brand
>
> **ShelveLoc:**   The quality of the shelve location (Bad, Good, Medium) for the product in the store
>
> **Age:**   The average age of the population for the region the store is located in
>
> **Education:**   The average education level (in years of schooling) for the store's region
>
> **Urban:**   Classification of the store's region as urban (Yes) or not urban (No)
>
> **US:**   Is the store located in the United States of America (Yes or No)
>
> **Validation:**   Randomly selected subsetting of the data into Training and Validation, to be used for cross-validation.
>
> *Note: The data set is from* An Introduction to Statistical Learning *by James and Witten, 2013.*

To fit generalized regression models to this data, we begin with the **Analyze > Fit Model** dialog (Figure 9.14), and specify the model **Personality** as **Generalized Regression**. The **Validation** column, specifying the training and validation portions of the data, is used in the validation role.

Figure 9.14: Generalized Regression Dialog (Fit Model)

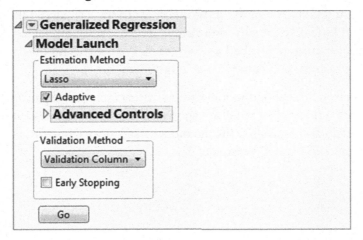

The resulting **Generalized Regression Model Launch** window is shown in Figure 9.15.

Figure 9.15: Generalized Regression Model Launch Window

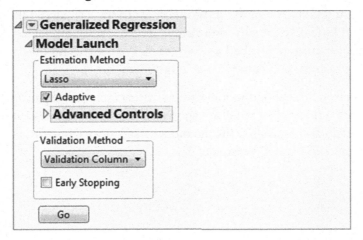

From this model launch window, we have a number of options:

- *Estimation Method*: The estimation method to use. Choices include Maximum Likelihood, Forward (Stepwise) Selection, Ridge Regression, Lasso, and Elastic Net.

- *Validation Method*: If a cross-validation column is used, **Validation Column** is the default. Other options are k-fold cross-validation, leave-one-out, a separate randomly chosen training/validation subset, AICc, or BIC.

- *Early Stopping*: If early stopping is selected, the model fitting process will stop adding complexity to the model when the chosen validation metric stops improving.

- *Adaptive*: This check box will appear and is checked by default when either **Lasso** or **Elastic Net** is chosen as the estimation method. It is recommended to use this JMP Pro default. Models derived using the adaptive Lasso and adaptive Elastic Net can have better statistical properties than models produced with the standard Lasso and Elastic Net algorithms.

- *Advanced Controls*: These options allow some additional control over the model fitting process; but, in general, you should not need to adjust these settings.

Maximum Likelihood Regression

The first model we fit is the standard linear regression model. This option is available as **Maximum Likelihood** under **Estimation Method**. This model fitting procedure does not use a regularization penalty, and should give the same model results as the **Fit Least Squares** platform (described in Chapter 4). We use this model as the baseline model against which we will compare our other models.

The maximum likelihood model results, which include the parameter estimates for each predictor on its original scale, are shown in Figure 9.16. Note that JMP Pro uses *centered and scaled predictors* when it fits the model. This means that, for a continuous predictor, X, the centered and scaled version of X is:

$$X^* = \frac{X - \overline{X}}{(n-1)S_X}$$

where $\overline{X}$ is the average value of X, n is the number of observations in the training data, and S_X is the standard deviation of X. This transformation ensures that the parameter estimates are all on the same relative scale, which allows the penalty functions to treat each parameter equally.

Figure 9.16:   Maximum Likelihood Regression Results

Maximum Likelihood with Validation Column Validation

Model Summary

Response	Sales
Distribution	Normal
Estimation Method	Maximum Likelihood
Validation Method	Validation Column
Mean Model Link	Identity
Scale Model Link	Identity

Measure	Training	Validation
Number of rows	189	211
Sum of Frequencies	189	211
-LogLikelihood	263.3082	311.92959
BIC	594.75911	693.43334
AICc	554.6964	651.7069
Generalized RSquare	0.8820374	0.8578891

Parameter Estimates for Original Predictors

Term	Estimate	Std Error	Wald ChiSquare	Prob > ChiSquare	Lower 95%	Upper 95%
Intercept	11.004227	0.8504904	167.40956	<.0001*	9.3372965	12.671158
CompPrice	0.0848644	0.0056941	222.12451	<.0001*	0.0737041	0.0960247
Income	0.0184102	0.0026878	46.915652	<.0001*	0.0131421	0.0236782
Advertising	0.1296904	0.0156897	68.32567	<.0001*	0.098939	0.1604417
Population	0.0001821	0.0004886	0.1388552	0.7094	-0.000776	0.0011397
Price	-0.091783	0.0037011	614.9847	<.0001*	-0.099037	-0.084529
ShelveLoc[Bad]	-4.828504	0.2235504	466.52426	<.0001*	-5.266655	-4.390353
ShelveLoc[Medium]	-2.944324	0.185831	251.0353	<.0001*	-3.308546	-2.580102
Age	-0.048141	0.0046078	109.15555	<.0001*	-0.057172	-0.03911
Education	-0.027803	0.027396	1.0299225	0.3102	-0.081498	0.0258923
Urban[No]	-0.279468	0.1606648	3.0256712	0.0820	-0.594365	0.0354295
US[No]	0.3898696	0.2071283	3.5429085	0.0598	-0.016094	0.7958335
Scale	0.9745559	0.0501257	378	<.0001*	0.8763112	1.0728005

Ridge Regression

To fit a ridge regression model, we choose **Ridge** from the **Estimation Method** and click **Go**. The results of the fitted model are shown in Figure 9.17.

Two **Solution Path** graphs are provided, along with other output. The horizontal axis on both graphs is labeled **Magnitude of the Scaled Parameter Estimates**, and this is the sum of the absolute values of the parameter estimates, based on the centered and scaled predictors. The vertical axis on the left-hand graph is the size of the individual scaled parameter estimates, and the connected lines on this chart represent the value of the parameter estimate for a particular predictor at different values of the penalty parameter.

The right-hand graph uses the Scaled –LogLikelihood, at different values of the penalty parameter (as described earlier in this chapter). By selecting and moving the vertical line on these plots, you tell JMP Pro to use a different penalty tuning parameter, which will lead to either more regularization (moving to the left) or less regularization (moving to the right).

Figure 9.17: Ridge Regression Results

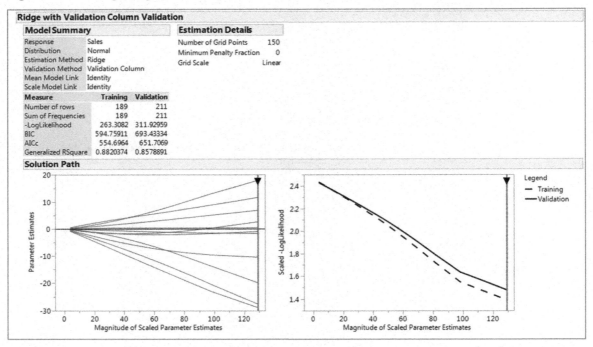

Parameter estimates are provided for the centered and scaled predictors, and the equivalent estimates are given for the original (untransformed) predictors (Figure 9.18). Note that these estimates, and all other reported statistics, update automatically when the penalty parameter is changed (by dragging either vertical line). We use the values that JMP Pro determines automatically based on cross-validation.

For this example, fitting a ridge regression model gives results identical to the maximum likelihood model. This is because the number of parameters in the model is much smaller than the number of rows in the table, and there are no linear dependencies between the predictor variables. Ridge regression is most beneficial when the number of predictors in the model is large, relative to the number of rows of data, or when there are strong correlations between the predictor variables.

Figure 9.18:   Ridge Regression Parameter Estimates

Parameter Estimates for Centered and Scaled Predictors

Term	Estimate	Std Error	Wald ChiSquare	Prob > ChiSquare	Lower 95%	Upper 95%
Intercept	7.6641799	0.0708885	11689.063	<.0001*	7.525241	7.8031188
CompPrice	17.830377	1.0937445	265.75978	<.0001*	15.686677	19.974077
Income	6.9148612	0.981428	49.64209	<.0001*	4.9912977	8.8384247
Advertising	11.655226	1.2998749	80.396712	<.0001*	9.1075179	14.202934
Population	0.3778916	0.9975432	0.1435063	0.7048	-1.577257	2.3330402
Price	-28.86273	1.1026778	685.13696	<.0001*	-31.02393	-26.70152
ShelveLoc[Bad]	-27.5972	1.1596679	566.32103	<.0001*	-29.87011	-25.32429
ShelveLoc[Medium]	-19.99922	1.144564	305.31334	<.0001*	-22.24252	-17.75591
Age	-10.4651	1.0643941	96.667742	<.0001*	-12.55127	-8.378922
Education	-1.012387	1.0318313	0.9626668	0.3265	-3.034739	1.0099649
Urban[No]	-1.745157	1.0000192	3.0454567	0.0810	-3.705159	0.2148444
US[No]	2.5723827	1.4253669	3.2570004	0.0711	-0.221285	5.3660506
Scale	0.9745559	0.05555	307.78383	<.0001*	0.8656799	1.0834318

Parameter Estimates for Original Predictors

Term	Estimate	Std Error	Wald ChiSquare	Prob > ChiSquare	Lower 95%	Upper 95%
Intercept	11.004227	0.98335	125.2284	<.0001*	9.0768964	12.931558
CompPrice	0.0848644	0.0052057	265.75978	<.0001*	0.0746614	0.0950674
Income	0.0184102	0.002613	49.64209	<.0001*	0.0132888	0.0235315
Advertising	0.1296904	0.014464	80.396712	<.0001*	0.1013414	0.1580393
Population	0.0001821	0.0004806	0.1435063	0.7048	-0.00076	0.001124
Price	-0.091783	0.0035065	685.13696	<.0001*	-0.098655	-0.08491
ShelveLoc[Bad]	-4.828504	0.2028996	566.32103	<.0001*	-5.22618	-4.430828
ShelveLoc[Medium]	-2.944324	0.168505	305.31334	<.0001*	-3.274588	-2.61406
Age	-0.048141	0.0048964	96.667742	<.0001*	-0.057738	-0.038544
Education	-0.027803	0.0283368	0.9626668	0.3265	-0.083342	0.0277363
Urban[No]	-0.279468	0.1601421	3.0454567	0.0810	-0.59334	0.034405
US[No]	0.3898696	0.2160282	3.2570004	0.0711	-0.033538	0.8132771
Scale	0.9745559	0.05555	307.78383	<.0001*	0.8656799	1.0834318

As an exercise, we have you attempt to fit a model to this data set with many interaction and quadratic terms. In this situation, ridge regression can be more useful.

Lasso Regression

We return to the model launch section of the output window, and choose **Lasso** as the estimation method. After the model is run, the model summary, solution path and model estimates are provided (Figures 9.19 and 9.20).

An interesting point is that, for the lasso model, not all original parameters are in the final model—some parameters are zeroed out in the parameter estimates tables.

Figure 9.19:   Lasso Regression Results

Adaptive Lasso with Validation Column Validation

Model Summary	
Response	Sales
Distribution	Normal
Estimation Method	Adaptive Lasso
Validation Method	Validation Column
Mean Model Link	Identity
Scale Model Link	Identity

Estimation Details	
Number of Grid Points	150
Minimum Penalty Fraction	0
Grid Scale	Linear

Measure	Training	Validation
Number of rows	189	211
Sum of Frequencies	189	211
-LogLikelihood	267.78439	306.20502
BIC	582.74451	660.57676
AICc	554.57438	631.30556
Generalized RSquare	0.8763154	0.8653947

Solution Path

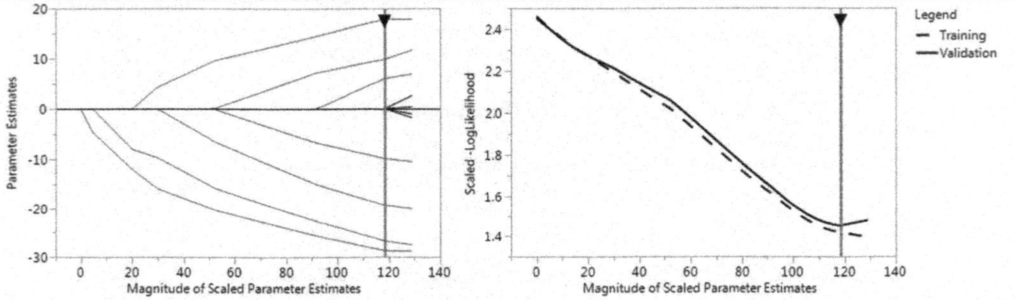

Figure 9.20:   Lasso Regression Parameter Estimates, Original Scale

Parameter Estimates for Original Predictors

Term	Estimate	Std Error	Wald ChiSquare	Prob > ChiSquare	Lower 95%	Upper 95%
Intercept	10.731622	0.7970059	181.30413	<.0001*	9.1695192	12.293725
CompPrice	0.0851007	0.005317	256.17027	<.0001*	0.0746795	0.0955218
Income	0.0159993	0.0026505	36.437029	<.0001*	0.0108044	0.0211942
Advertising	0.1095912	0.0101097	117.51102	<.0001*	0.0897766	0.1294057
Population	0	0	0	1.0000	0	0
Price	-0.091521	0.0035922	649.10536	<.0001*	-0.098561	-0.08448
ShelveLoc[Bad]	-4.691478	0.2012714	543.31958	<.0001*	-5.085963	-4.296994
ShelveLoc[Medium]	-2.840111	0.1726371	270.64661	<.0001*	-3.178474	-2.501749
Age	-0.045345	0.0050421	80.880225	<.0001*	-0.055227	-0.035463
Education	0	0	0	1.0000	0	0
Urban[No]	0	0	0	1.0000	0	0
US[No]	0	0	0	1.0000	0	0
Scale	0.9979123	0.0569093	307.48109	<.0001*	0.8863721	1.1094525

Here, we see that the results generated using lasso are different from the results produced using ridge regression. The validation RSquare (under **Generalized RSquare**) for this model is slightly higher, so this model is doing slightly better than the Ridge

Regression model at making predictions. Also, four of the parameter estimates in this model are zero, and all of the other parameter estimates are either very close to or smaller than the estimates provided by the ridge regression model. The lasso method has *shrunk* many of these parameter estimates, providing a reasonable model with less complexity than ridge or maximum likelihood.

Elastic Net

In the **Model Launch** dialog we now select **Elastic Net** as the estimation method. The fitted elastic net model results are shown in Figures 9.21 and 9.22.

The elastic net model has also produced a simpler regression model than the maximum likelihood or the ridge regression model. This model is very similar to the lasso model.

Figure 9.21:   Elastic Net Regression Results

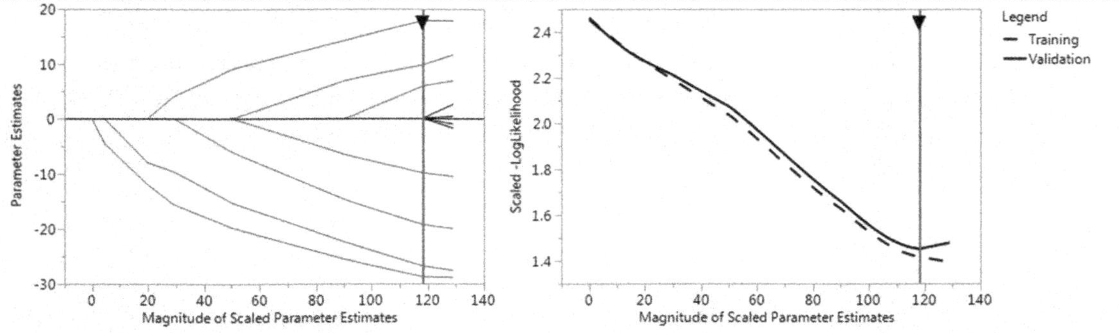

Figure 9.22:   Elastic Net Parameter Estimates

Parameter Estimates for Original Predictors						
Term	Estimate	Std Error	Wald ChiSquare	Prob > ChiSquare	Lower 95%	Upper 95%
Intercept	10.739643	0.797401	181.39537	<.0001*	9.1767661	12.30252
CompPrice	0.0849691	0.0053243	254.68142	<.0001*	0.0745337	0.0954045
Income	0.0158629	0.0026565	35.655804	<.0001*	0.0106562	0.0210697
Advertising	0.1093711	0.0101187	116.83132	<.0001*	0.0895389	0.1292033
Population	0	0	0	1.0000	0	0
Price	-0.091444	0.0035983	645.81278	<.0001*	-0.098496	-0.084391
ShelveLoc[Bad]	-4.685432	0.2014764	540.81805	<.0001*	-5.080319	-4.290546
ShelveLoc[Medium]	-2.8349	0.1728984	268.83997	<.0001*	-3.173775	-2.496026
Age	-0.045226	0.0050432	80.42202	<.0001*	-0.055111	-0.035342
Education	0	0	0	1.0000	0	0
Urban[No]	0	0	0	1.0000	0	0
US[No]	0	0	0	1.0000	0	0
Scale	0.9982908	0.0570647	306.04035	<.0001*	0.886446	1.1101356

As we saw in Chapter 8, we can save the prediction formulas to the data table and formally compare these models using the **Model Comparison** platform. One advantage of the regularized (penalized) regression models over decision trees and neural network models is that the saved prediction formulas have the same general structure as formulas saved for regression models. This makes it easier to interpret the models produced using the **Generalized Regression** platform.

Other Applications of Generalized Regression

In the example above, we illustrated how to use the **Generalized Regression** platform in JMP Pro to approach a standard linear regression problem. There are other nonstandard situations where this platform can be applied. When **Generalized Regression** is selected as the **Personality** in the **Fit Model Dialog**, the **Distribution** option is available (see Figure 9.23). This setting specifies the "error" distribution that is used for building the model.

Figure 9.23: Generalized Regression Distributions

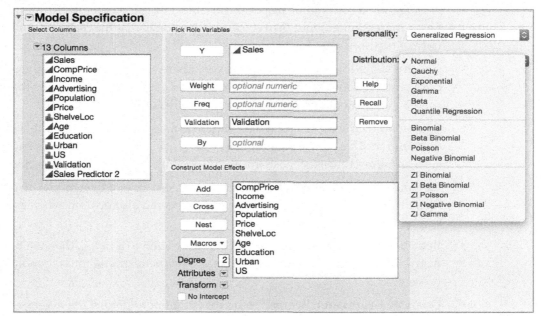

In Chapter 4, "Multiple Linear Regression," we discussed the assumption that the distribution of the residuals follows a normal distribution. In many situations, the residuals won't follow a normal distribution, may have a high percentage of outliers, or may be constrained to be within a range of values. For categorical response models, we encounter similar issues. In Chapter 5, we discussed the logistic regression model for predicting a binary (two-level) response. What we didn't mention is that the underlying "error" model that is assumed for logistic regression is the binomial distribution. There are situations for categorical response models where we would expect the errors to follow other distributions, such as when the response is a count of the number events in a given time period. Ultimately, the choice of the error distribution specifies which likelihood function is used to estimate the parameters in the model.

The available error distributions are listed below, along with a brief description of when they can (or should) be used:

Continuous Response Models

- **Normal**: For a continuous response variable, this is the default distribution. For linear regression models, the typical assumption is that the residuals will have a normal distribution.

- **Cauchy**: For a continuous response, when there are "heavy tails" in the residual error distribution. This distribution allows for a model fitting process that is robust to having many outliers in the model residuals.

- **Exponential**: The exponential distribution is a positive valued, "right-skewed" distribution. This can be useful if the response is related to time-to-event data, such as failure times for mechanical components or times between accidents.

- **Gamma**: The exponential distribution is a special case of the gamma distribution, as is the Weibull distribution. This distribution can also be used when the response is related to time-to-event data.

- **Beta**: If the response is a proportion or constrained to a range, then the beta distribution can incorporate this constraint on the response into the modeling process.

- **Quantile Regression**: Regression models attempt to predict the average value of the response, conditional on the model predictors. If, instead, you want to predict a percentile (or quantile) of the response, you can use this option. An example would be if the response of interest is the amount of an insurance claim. Quantile regression can be used to estimate the 90th percentile of the insurance claims, conditional on insurance risk factors.

Categorical Response

- **Binomial**: If the response is a binary (0/1, Yes/No, …) variable, then this is an appropriate choice. This produces a model that is similar to a logistic regression model. If the response is related to the number or proportion of successes or failures out of a known number of trials, it may also be appropriate to use this distribution (but you must specify the number trials).

- **Beta-Binomial**: This is similar to the binomial distribution for the multiple trials case, but it allows for broader variation in the residual distribution.

- **Poisson**: If the response is counts of the number of events that occurred in a fixed time period, this residual distribution could be appropriate. An example of this type of data is the number of telephone calls per day received at a customer call center. This is also known as Poisson regression.

- **Negative Binomial**: Also used if the response is counts of events, but when the response is based on the number of successes before a specified number of failures. This is useful if the count data aren't necessarily collected over a fixed time period, or if the events happen at multiple locations with different rates of occurrence of events.

- **Zero-Inflated Distributions**: A zero inflated distribution is a mixture of distributions, where the "Zero Inflation" is the proportion of the time that the outcome is exactly zero, but otherwise follows a specified type of distribution. An example is when the process that is being recorded has some threshold to overcome before events can be counted or measured. Zero inflated binomial, beta binomial, Poisson, negative binomial, or gamma error distributions are available in the Generalized Regression platform.

The real power of the **Generalized Regression** platform is that, for all of the special cases listed above, you can also use regularization methods like the lasso and elastic net to estimate the model parameters for the specified model. A practical example of this is using the lasso regularization method to choose the appropriate parameters in a logistic regression (using the binomial error distribution), rather than doing stepwise regression, as shown in Chapter 5.

Key Take-Aways

In Chapters 3 through 8, we described modeling methods that are considered "core methods" for any analyst to have in their toolset. These core methods are available in the standard version of JMP (along with JMP Pro). The methods described in this chapter have been developed over the last few decades as enhancements to these core tools. We have called these "advanced" models, because all of these methods are more computationally challenging, and the underlying mathematics are more complex. However, modern software tools like JMP Pro enable you to easily fit these models and to apply them to solve your real-world analytical problems.

Exercises

Exercise 9.1: Use the **Titanic Passengers BBM Ch9.jmp** data set for this exercise. Recall that the response is **Survived** (we introduced this example in Chapter 5).

a. Fit a bootstrap forest model using the default settings. Use model cross-validation (use the Validation column in the data set). Set the random seed to 1000 prior to fitting the model (for repeatability).

b. Turn on **Column Contributions** for this model. Which predictors are explaining the most about the probability that a passenger would have survived?

c. Save the prediction formula to the data table, and use the **Profiler** (from the **Graph** menu) to graphically examine the probability formulas. For each of the variables that have the largest contributions, what is the effect of these variables on the predicted probability of surviving?

d. Repeat steps a through c using a boosted tree model.

e. Compare the performance of the models produced above using the **Model Comparison** platform.

   i. Which criteria should be used for comparing these models?

   ii. Which model was the best performer? Explain.

Exercise 9.2: Repeat Exercise 9.1, but use the data set **Churn BBM Ch9.jmp** to predict the response **target churn** as a function of the 14 variables grouped together as predictors.

Exercise 9.3: For the data set **Car Seats BBM Ch9.jmp**, described earlier in this chapter, fit and compare the following models (use the Validation column in the data table for cross-validation):

a. Bootstrap Forest

b. Boosted Tree

c. Boosted Neural Network

d. Generalized Regression using Ridge, Lasso, and Elastic Net on a response surface model for the predictors.

   Include all of the main effects, two-factor interactions, and quadratic (squared) terms for each predictor in the regularized models. (To do this, select the model factors in the **Fit Model** dialog, click **Macro**, and choose **Response Surface**, as shown in Figure 9.24.)

Figure 9.24:   Adding Interactions and Quadratic Effects to a Model

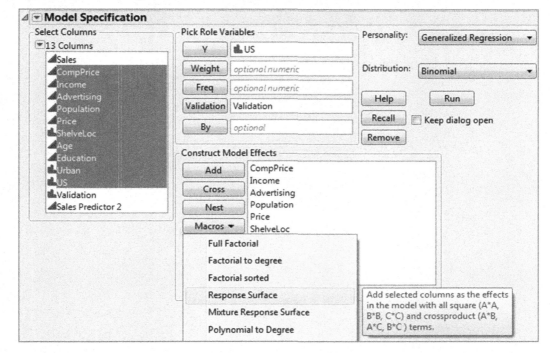

Exercise 9.4: Use the **Churn BBM Ch9.jmp** data set for this exercise.

a.   Fit **Generalized Regression** models to predict churn (using the *Binomial* error distribution). Use the Lasso and Elastic Net estimation methods and cross-validation.

b.   Compare these regularized models, and select the best model. Which model performed better on the validation set?

c.   Compare all four models produced using the Model Comparison platform.

   i.   Which criteria should be used for comparing these models?

   ii.   Which model was the best performer? Explain.

References

James, Gareth, Daniela Witten, Trevor Hastie, and Robert Tibshirani. 2013. *An Introduction to Statistical Learning*. Springer.

JMP Documentation. 2015. *JMP 12 Specialized Models*. SAS Institute Inc.

10

Capstone and New Case Studies

Introduction

In this chapter, we present four case studies. The first, **Cell Classification**, is the only comprehensive example, or capstone, illustrating the entire Business Analytics Process but in a non-business environment. The second, **Bulldozer Blue Book**, is based on a Kaggle contest to predict the selling price of heavy equipment. The third and fourth case studies are only briefly introduced, leaving the analysis and modeling as exercises: **Default Credit Card** provides an outline for presenting results to management; and **Carvana**, which is also based on a Kaggle contest, requires importing a data set from the Internet. For all four case studies, we provide a series of exercises at the end of the chapter to provide an opportunity to apply what you have learned throughout this book and to tie the analytic concepts and tools together.

Since we generally refer to methods introduced previously, for the most part we omit the JMP keystrokes and details in this chapter.

Case Study 1:   Cell Classification

In the following case study, we walk through the stages of the Business Analytics Process, introduced in Chapter 2.   Rather than addressing a business problem, the focus of this case study is a scientific research problem. As we will illustrate, the BAP is broadly applicable to any field that uses predictive modeling, such as marketing, medicine, manufacturing, and the sciences.

The objective of this research project is to predict whether a breast lump, or tumor, is malignant or benign. We apply only a handful of the exploratory and modeling tools covered in this book. We ask you to conduct a more thorough analysis in an exercise.

Stage 1:   Define the Problem

The key activities in this important first stage are:

- Understanding the business problem (or simply, the problem), project objectives, and importance to the organization.
- Framing the analytics problem(s)
- Defining the project goal and time frame
- Developing a project plan and time line
- Obtaining resources and approval to start project

Here, we focus on the problem, the project objective, and the analytics problem, and omit the other details for the sake of brevity. Since this is a research-oriented project, we define the project in terms of the research problem instead of the business problem.

The Research Problem and the Objective

Breast tumors may be malignant or benign. Current methods for making diagnoses (for example, radiology) have both a high false positive rate and a high false negative rate. That is, the tests often indicate that a tumor is malignant when it isn't, or conclude the tumor is benign when it is, in fact, malignant.

Researchers are exploring an alternative approach involving digital images of the nuclei of cells in fluid removed from tumors. Can knowledge of certain characteristics of these nuclei provide an accurate prediction of malignancy?

The Analytics Problem

We use the data available to develop a useful prediction model for malignancy. Since we're interested in the ability of our final model to correctly classify a tumor, we define "useful" as having a low misclassification rate. False positives and false negatives are the two potential types of model misclassification.

Stage 2: Prepare for Modeling

Recall that the key activities in this stage are:

- Collecting, cleaning, and transforming the data
- Defining relevant features in the data
- Examining and understanding the data
- Producing data sets that are ready for analysis and model-building

We use a publicly available data set, obtained from a study conducted at the University of Wisconsin (Mangasarian, Street, & Wolberg, 1995).

The Data   CellClassification BBM.jmp

The data set contains a total of 569 records, of which 357 tumors are benign and 212 are malignant. Each tumor contained multiple cells, and image analysis was used to determine the following ten features for each cell:

> **Radius:**   Mean of distances from the center to points on the perimeter
>
> **Texture:**   Standard deviation of gray-scale values
>
> **Perimeter:**   Perimeter of the cell nucleus

Area: Area of the cell nucleus

Smoothness: Local variation in radius lengths

Compactness: Perimeter/area

Concavity: Severity of concave portions of the contour

Concave Points: Number of concave portions of the contour

Symmetry: Symmetry of the cell nucleus

Fractal dimension: Regularity of the boundary (the contour) of the nucleus

For each tumor image, the mean and standard deviation of the feature values, for all of the cells in the image, were calculated (labeled with the prefix "Mean" or "SE", respectively, in the column name). Also, the average of the 3 largest values for each feature in the tumor image was calculated (labeled with the prefix "Max" in the column name). This resulted in thirty potential predictor variables that could be useful for classifying tumors as benign or malignant.

Examine and Understand the Data

As we have seen, several data visualization techniques can be used to look at individual variables and explore the relationships between predictors and between the predictors and the response. We use histograms (with their inherent dynamic linking) and scatterplot matrices. However, the other interactive visual tools available in JMP could also be used to explore the data. The goal is to become familiar with the variables and identify any potential data quality issues or challenges.

In Figure 10.1, we see the distribution of the response variable (**Diagnosis**) and three of the potential predictors (**Mean Radius**, **Mean Perimeter**, and **Mean Area**). The bar for the **Malignant** category of the **Diagnosis** distribution is selected, and the shaded regions in the other graphs show the distribution of the three predictors for the malignant tumors. It appears that **Malignant** cells tend to have larger values for all three of the predictors shown.

Figure 10.1: Distributions of Outcome and Potential Predictors

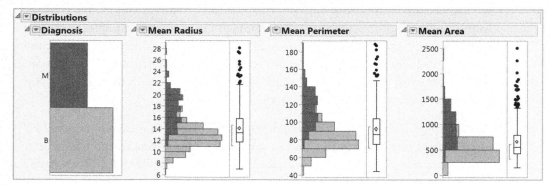

As we have discussed in the earlier chapters, it is a good practice to examine the distribution of the response and all of the potential predictors, but for the sake of brevity, we omit doing this here. Fully exploring the general characteristics of the data makes it easier to detect potential problems (such as data errors or outliers) that could impact the analysis.

It is also a good practice to explore bivariate relationships between the response and the potential predictors using scatterplots or box plots. In the **Graph Builder** output in Figure 10.2, we can see that the **Mean Area** of cells is very different for the two **Diagnosis** groups, but the differences aren't as large for **Mean Texture**.

Figure 10.2: Box Plots of Diagnosis and Mean Area and Mean Texture

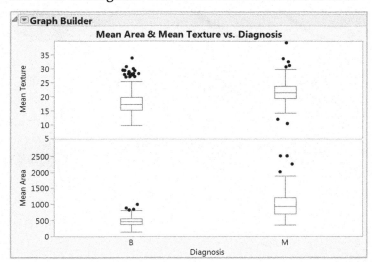

In addition to exploring potential relationships between the response and the predictors, we explore the relationship between the potential predictor values. We could construct, individually, all of the two-way scatterplots of the predictor variables, resulting in 450 separate graphs. A simpler way to see all of these relationships is to use a scatterplot matrix (available in the JMP menu under **Graph > Scatterplot Matrix**) . A scatterplot matrix shows all of the two-way graphs arranged in a grid. In this example, the scatterplot matrix of the 30 predictors is a 30 x 30 matrix.

A scatterplot matrix of the first five potential predictors is shown in Figure 10.3. These graphs show a strong positive (but in some cases non-linear) relationship between the first three pairs of predictors (**Mean Radius**, **Mean Perimeter**, and **Mean Area**), and a weaker relationship amongst the other pairs.

Figure 10.3:   Scatterplot Matrix of the First Five Variables

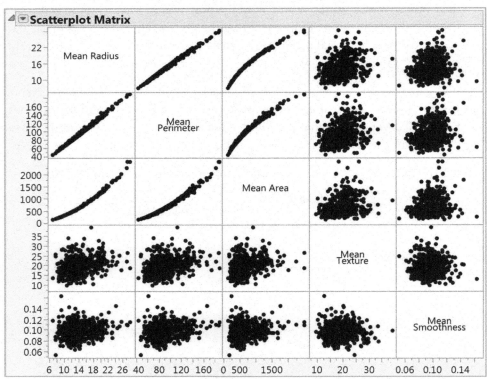

By using **Diagnosis** as a legend, where the diagnosis groups have different markers and colors (see Figure 10.4), we can further explore potential relationship between these five predictors and **Diagnosis**.

Figure 10.4:   Scatterplot Matrix with Legend and Markers

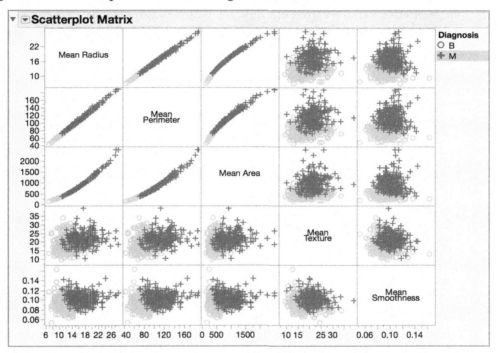

Malignant cells (highlighted and darkened) appear to have larger and possibly more extreme values for the variables shown than Benign cells.

Many of the 30 potential predictors show a strong relationship with one other (not shown here). This correlation between predictors (multicollinearity) can be problematic when building predictive models (for reasons discussed in Chapter 4).

One approach to address these problems is to reduce the "dimensionality" of our data. In Chapter 3, we briefly introduced this subject, and introduced some popular dimension-reduction methods. These tools can be used to reduce the number of potential predictor variables under consideration before building the model. One such method is *variable clustering*, which finds groups of variables that are similar to one another, and then selects a representative variable from each group. (Recall that Variable Clustering is available in the Principal Components platform in JMP Pro from the top red triangle - the option is **Cluster Variables**).

For our data, variable clustering (Figure 10.5) provides us with the six most representative variables from our data set, which account for nearly 80% of the variation in the predictor set. For instance, **Mean Area** is the most representative variable of a group of 10 predictor variables.

Figure 10.5:   Variable Clustering with Most Representative Variables

◢ ▾ **Variable Clustering**

◢ **Cluster Summary**

Cluster	Number of Members	Most Representative Variable	Cluster Proportion of Variation Explained	Total Proportion of Variation Explained	.2 .4 .6 .8
1	10	Mean Area	0.856	0.285	
2	6	Max Concavity	0.817	0.163	
3	5	SE Fractal Dim	0.726	0.121	
6	4	Mean Smoothness	0.673	0.09	
4	2	Mean Texture	0.956	0.064	
5	3	SE Symmetry	0.605	0.06	

Proportion of variation explained by clustering: 0.784

A second dimension-reduction method involves using decision trees (or other tree-based methods such as bootstrap forest) to identify predictor variables that contribute most to distinguishing between a "malignant" response and a "benign" response (see Chapters 6 and 9 for more details on tree-based methods).   The predictors identified using a decision tree (Figure 10.6), in descending order of influence, are **Max Area**, **Max Concave Points**, **Mean Concavity**, **Mean Texture**, and **Max Texture**.

Figure 10.6:   Column Contributions of Top Five Variables

◢ **Column Contributions**

Term	Number of Splits	G^2		Portion
Max Area	1	441.855907		0.6786
Max Concave Points	2	119.587459		0.1837
Mean Concavity	1	39.8618559		0.0612
Max Texture	1	30.2731361		0.0465
Mean Texture	1	19.5849428		0.0301

Combining the lists of predictor variables from both of these dimension reduction methods yields a reduced set of candidate variables consisting of nine members: **Mean Area, Mean Texture, Mean Concavity, Max Area, Max Texture, Max Smoothness, Max Concave Points, SE Compactness** and **SE Symmetry**.

In this step of the analytics process, we have passed over other important issues, again for the sake of brevity, such as assessing the cleanliness of the data (missing values, erroneous data, outliers, etc.) and determining whether data transformations are required (refer to Chapter 3 for these details). However, we leave Stage 2 with some insight into our data and a reduced set of predictor variables to consider as we enter the modeling phase of the project. (Note that, in practice, all available predictors may be included in the model, and other effects, such as interactions, might be considered.)

Stage 3:   Modeling

In this step, the team:

- Chooses the appropriate modeling methods
- Fits one or more models
- Evaluates the performance of each model
- Chooses the best model or set of models to address the analytics problem (and ultimately the business problem)

The data set contains a validation column, which uses a formula to partition our data into the training, validation, and test sets. We use roughly 50% of the data to train our models, 30% for validation, and the remaining 20% to test and assess the performance of the selected model (Figure 10.7).

Figure 10.7:   Training, Validation and Test Sets

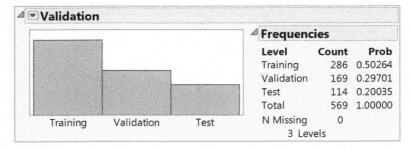

This example involves a categorical response, **Diagnosis**. We decide to fit four types of predictive models: logistic regression, bootstrap forest, boosted tree, and neural network. We briefly describe these models, but omit the details discussed in prior chapters. (Note that, in practice, additional and more complex models would be built. We fit these four basic models, at the default settings, for illustration purposes.)

Recall that our primary interest is in developing a model that correctly predicts malignancy. So our primary criterion in comparing the performance of the models that we build is the misclassification rate: that is, the proportion of observations misclassified by our model.

Model 1: Stepwise Logistic Regression

A stepwise regression logistic regression model is fit using the reduced set of nine predictors (for details on logistic regression, see Chapter 5). The default stopping rule, Max Validation RSquare, retains all nine predictors. Since many of these predictors are not significant, we also apply Minimum AICc and Minimum BIC, and settle on a reduced model with four predictors (Figure 10.8).

Figure 10.8: Reduced Logistic Regression Model (4 predictors)

◢ Parameter Estimates				
Term	**Estimate**	**Std Error**	**ChiSquare**	**Prob>ChiSq**
Intercept	22.3422337	5.0759749	19.37	<.0001*
Mean Area	0.01765505	0.0090637	3.79	0.0514
Max Area	-0.0256123	0.0084437	9.20	0.0024*
Max Texture	-0.1997831	0.0793884	6.33	0.0119*
Max Concave Points	-57.170596	16.527206	11.97	0.0005*

The misclassification rate for the training data is 2.1%. That is, only 2.1% of the observations in the training set were misclassified. The **Confusion Matrix** for the training data (bottom left, in Figure 10.9) indicates that four malignant lumps were classified as benign, while two benign lumps were classified as malignant.

Figure 10.9: **Misclassification Rates and Confusion Matrix**

▼ **Whole Model Test**

Model	-LogLikelihood	DF	ChiSquare	Prob>ChiSq
Difference	163.99335	4	327.9867	<.0001*
Full	20.48613			
Reduced	184.47948			

RSquare (U)	0.8890
AICc	51.1866
BIC	69.2522
Observations (or Sum Wgts)	286

Measure	Training	Validation	Test	Definition
Entropy RSquare	0.8890	0.8122	0.8924	1-Loglike(model)/Loglike(0)
Generalized RSquare	0.9415	0.8994	0.9445	$(1-(L(0)/L(model))^{(2/n)})/(1-L(0)^{(2/n)})$
Mean -Log p	0.0716	0.1274	0.0713	$\sum$ -Log($\rho[j]$)/n
RMSE	0.1310	0.1660	0.1447	$\sqrt{\sum(y[j]-\rho[j])^2/n}$
Mean Abs Dev	0.0388	0.0492	0.0498	$\sum$ |y[j]-$\rho[j]$|/n
Misclassification Rate	0.0210	0.0473	0.0263	$\sum$ ($\rho[j]\neq\rho$Max)/n
N	286	169	114	n

▶ **Lack Of Fit**

▶ **Parameter Estimates**

▶ **Effect Likelihood Ratio Tests**

▼ **Confusion Matrix**

Training			Validation			Test		
Actual	**Predicted**		**Actual**	**Predicted**		**Actual**	**Predicted**	
Diagnosis	**B**	**M**	**Diagnosis**	**B**	**M**	**Diagnosis**	**B**	**M**
B	185	2	B	97	2	B	70	1
M	4	95	M	6	64	M	2	41

The misclassification rates for the validation and test sets are slightly higher (middle, Figure 10.9). This is not unexpected, since these observations were not used in the model building process.

Model 2:   Bootstrap Forest

A bootstrap forest is grown by averaging the predicted values of many decision trees built on bootstrap samples of the training data. In each tree, predictor variables are randomly sampled for inclusion in the model fitting (see Chapter 9).

The bootstrap forest results in Figure 10.10 show a 2.45% misclassification rate for the training data. Again, we see higher misclassification rates for the hold-out samples (7.1% for the validation set and 7.02% for the test set).

Figure 10.10:   Bootstrap Forest Model Results

▼ ⊟**Bootstrap Forest for Diagnosis**
  ▶ **Specifications**
  ▼ **Overall Statistics**

Measure	Training	Validation	Test	Definition
Entropy RSquare	0.8240	0.7306	0.7416	1-Loglike(model)/Loglike(0)
Generalized RSquare	0.9032	0.8469	0.8522	$(1-(L(0)/L(\text{model}))^{(2/n)})/(1-L(0)^{(2/n)})$
Mean -Log p	0.1136	0.1828	0.1713	$\sum -\text{Log}(\rho[j])/n$
RMSE	0.1597	0.2280	0.2170	$\sqrt{\sum(y[j]-\rho[j])^2/n}$
Mean Abs Dev	0.0905	0.1304	0.1341	$\sum |y[j]-\rho[j]|/n$
Misclassification Rate	0.0245	0.0710	0.0702	$\sum (\rho[j]\neq\rho\text{Max})/n$
N	286	169	114	n

  ▼ **Confusion Matrix**

Training			Validation			Test		
Actual	**Predicted**		**Actual**	**Predicted**		**Actual**	**Predicted**	
Diagnosis	**B**	**M**	**Diagnosis**	**B**	**M**	**Diagnosis**	**B**	**M**
B	184	3	B	96	3	B	68	3
M	4	95	M	9	61	M	5	38

Since the Bootstrap Forest involves random sampling of model terms and observations, your results may be slightly different if you try to replicate the results. To obtain the same results each time Bootstrap Forest is run with this data (and these variables), set the random seed to 2000 before launching the **Partition** platform. As discussed previously, the random seed can be set using use the **Random Seed Reset** add-in, available from the **JMP File Exchange**.

Model 3:   Boosted Tree

A boosted tree is a decision tree built by creating a sequence of small decision trees, where each tree predicts the scaled residuals from the previous decision tree. These smaller trees are then combined to form one larger tree (see Chapter 9 for details).

The boosted tree results in Figure 10.11 show a misclassification rate of 0.0035 (or, 0.35%) for the training data. However, as we have seen, the misclassification rates for the validation and test sets are substantially higher (7.1% and 6.1% respectively). Again, set the random seed to 2000 before launching the **Partition** platform to obtain the results shown here.

Figure 10.11:   Boosted Tree Results

▼ ⊡ Boosted Tree for Diagnosis

▶ Specifications

▼ Overall Statistics

Measure	Training	Validation	Test	Definition
Entropy RSquare	0.9693	0.7324	0.8007	1-Loglike(model)/Loglike(0)
Generalized RSquare	0.9847	0.8482	0.8906	$(1-(L(0)/L(model))^{\wedge}(2/n))/(1-L(0)^{\wedge}(2/n))$
Mean -Log p	0.0198	0.1815	0.1321	$\sum -Log(p[j])/n$
RMSE	0.0521	0.2210	0.2086	$\sqrt{\sum(y[j]-p[j])^2/n}$
Mean Abs Dev	0.0180	0.0779	0.0792	$\sum |y[j]-p[j]|/n$
Misclassification Rate	0.0035	0.0710	0.0614	$\sum (p[j] \neq pMax)/n$
N	286	169	114	n

▼ Confusion Matrix

Training			Validation			Test		
Actual	**Predicted**		**Actual**	**Predicted**		**Actual**	**Predicted**	
Diagnosis	**B**	**M**	**Diagnosis**	**B**	**M**	**Diagnosis**	**B**	**M**
B	187	0	B	95	4	B	67	4
M	1	98	M	8	62	M	3	40

Model 4:   Neural Network

A neural network is a complex and generally non-linear model made up of inputs, hidden nodes that transform the inputs, and predicted outputs (see Chapter 7 for details). Neural models, by their very nature, tend to perform extremely well in predictive situations.

The resulting neural model has a misclassification rate of 1.75% for the training set, 3.55% for the validation set, and 2.63% for the test set (see Figure 10.12). As with Bootstrap Forest, there is a random component to neural networks, and to obtain the same results as shown in this case study, set the random seed to 2000.

In this example, we fit the default neural model. The model has one hidden layer with three nodes. At each node, the Sigmoid TanH function is used. However, since our goal is to produce a model with the best predictive ability, in practice we would produce several models, some with far more complexity, and select the best neural model. Additional options for fitting neural models are described in Chapters 7 and 9.

Figure 10.12: Neural Network Model Results

▼ ⊡**Neural**
Validation Column: Validation
▶ **Model Launch**
▼ ⊡**Model NTanH(3)**

▼ **Training**		▼ **Validation**		▼ **Test**	
▼ **Diagnosis**		▼ **Diagnosis**		▼ **Diagnosis**	

Measures	Value	Measures	Value	Measures	Value
Generalized RSquare	0.9444644	Generalized RSquare	0.9145167	Generalized RSquare	0.9597596
Entropy RSquare	0.8942105	Entropy RSquare	0.8376038	Entropy RSquare	0.9204403
RMSE	0.1258036	RMSE	0.1655919	RMSE	0.1211944
Mean Abs Dev	0.0344874	Mean Abs Dev	0.0508975	Mean Abs Dev	0.0409
Misclassification Rate	0.0174825	Misclassification Rate	0.035503	Misclassification Rate	0.0263158
-LogLikelihood	19.515986	-LogLikelihood	18.617314	-LogLikelihood	6.0103152
Sum Freq	286	Sum Freq	169	Sum Freq	114

Confusion Matrix

Actual Diagnosis	Predicted B	M	Actual Diagnosis	Predicted B	M	Actual Diagnosis	Predicted B	M
B	186	1	B	96	3	B	68	3
M	4	95	M	3	67	M	0	43

Confusion Rates

Actual Diagnosis	Predicted Rate B	M	Actual Diagnosis	Predicted Rate B	M	Actual Diagnosis	Predicted Rate B	M
B	0.995	0.005	B	0.970	0.030	B	0.958	0.042
M	0.040	0.960	M	0.043	0.957	M	0.000	1.000

Selecting and Testing the Model

When selecting the best model, we compare model performance on a data set not used to develop the model. As we have seen, the validation data is used behind the scenes to help determine the best model complexity for each model type. Now we use the test data for comparing competing models because it was completely held out of the model building process. Having a test set allows for us to make a fair assessment of each model's performance. For additional discussion of the importance of validation, types of validation and construction of the validation column, see Chapter 8.

We use the **Model Comparison** platform to compare these competing models (under **Analyze > Modeling**, with the **Validation** column in the **Group** or **By** field). Remember that, for each model, the prediction formulas must first be saved to the data table.

The Neural Network performs the best on the validation data (top in Figure 10.13), with a validation misclassification rate of only 3.55%. The logistic regression model is a close second best at 4.73%. These two models perform identically on the test data, with misclassification rates of 2.63%. (Recall that other options for comparing the models are available under the red triangles. For models with categorical responses, these options, including ROC and Lift Curves, were described in Chapter 6.)

Figure 10.13: Model Comparison – Validation (Top), Test (Bottom)

▼ **Model Comparison Validation=Validation**

▶ **Predictors**

▼ **Measures of Fit for Diagnosis**

Creator	.2 .4 .6 .8	Entropy RSquare	Generalized RSquare	Mean -Log p	RMSE	Mean Abs Dev	Misclassification Rate	N
Fit Ordinal Logistic		0.8122	0.8994	0.1274	0.1660	0.0492	0.0473	169
Bootstrap Forest		0.7306	0.8469	0.1828	0.2280	0.1304	0.0710	169
Boosted Tree		0.7324	0.8482	0.1815	0.2210	0.0779	0.0710	169
Neural		0.8376	0.9145	0.1102	0.1656	0.0509	0.0355	169

▼ **Model Comparison Validation=Test**

▶ **Predictors**

▼ **Measures of Fit for Diagnosis**

Creator	.2 .4 .6 .8	Entropy RSquare	Generalized RSquare	Mean -Log p	RMSE	Mean Abs Dev	Misclassification Rate	N
Fit Ordinal Logistic		0.8924	0.9445	0.0713	0.1447	0.0498	0.0263	114
Bootstrap Forest		0.7416	0.8522	0.1713	0.2170	0.1341	0.0702	114
Boosted Tree		0.8007	0.8906	0.1321	0.2086	0.0792	0.0614	114
Neural		0.9204	0.9598	0.0527	0.1212	0.0409	0.0263	114

Because the logistic regression model has a simpler mathematical form and is easier to interpret than the neural model, we apply the principle of parsimony and choose the logistic model as the best and most practical model to deploy. The terms and parameter estimates for the logistic regression model are shown in Figure 10.14.

Figure 10.14: Logistic Regression Model Terms

◢ **Parameter Estimates**

Term	Estimate	Std Error	ChiSquare	Prob>ChiSq
Intercept	22.3422337	5.0759749	19.37	<.0001*
Mean Area	0.01765505	0.0090637	3.79	0.0514
Max Area	-0.0256123	0.0084437	9.20	0.0024*
Max Texture	-0.1997831	0.0793884	6.33	0.0119*
Max Concave Points	-57.170596	16.527206	11.97	0.0005*

For log odds of B/M

The estimates provided for each term can be expressed in a formula to predict if a lump is benign or malignant. The formula has four predictors: **Mean Area**, **Max Area**, **Max Texture**, and **Max Concave Points**. Measurements of these four characteristics can be used to predict the probability that a lump is malignant.

Stage 4:   Deploy Model

Once a model has been developed (and perhaps after some additional testing or piloting to verify model performance), it is time to put our model into practice. This model will be applied to aid diagnosticians in determining whether biopsied breast cells are cancerous.

The model itself predicts the probability that the cells are Benign. The parameter estimates from the fitted model are used, and this prediction equation has the form:

$$P(Benign) = 1 / [1 + \exp\{22.3422 + 0.01766(Mean\ Area) - 0.0256(Max\ Area)$$

$$-0.1988(Max\ Texture) - 57.1706(Max\ Concave\ Points)\}]$$

Using the result from this equation, a decision is made as to whether cells are cancerous or not. If the result is greater than some threshold, T, it is classified as malignant (M); otherwise, it is classified as benign (B). The default value for T is 0.5, but alternate values can be examined. This can be done by examining how well this rule does at classifying the outcome for different values of T. In this case, it is more desirable that this test be sensitive toward detecting cancer at the risk of incorrectly classifying some individuals without cancer as actually having the disease. This makes sense, because it is much more serious to miss a diagnosis of cancer when in fact cancer is present.

Letting T be equal to 0.25, 0.50, and 0.75, the resulting model classifications are compared to the correct diagnosis. Again, the test data set is used to make this assessment. (Note that the **Alternate Cut-off Confusion Matrix Add-In**, on the **JMP File Exchange**, can be used to compare misclassification rates at different thresholds.)

As we can see in Figure 10.15, a threshold of 0.25 incorrectly classifies roughly 0.3% of malignant tumors as benign (the solid line is the misclassification rate for malignant tumors at different thresholds). However, at thresholds of 0.50 and 0.75, the model misclassifies roughly 0.6% of malignant tumors as benign. Because we are most concerned about a misclassification of a malignant diagnosis, we choose the more sensitive threshold of 0.25.

Note that, in this example, we select three thresholds for simplicity. In practice, additional thresholds should be explored to minimize the misclassification rate for the target category. In addition, the cost of misclassifications should be considered. This can be accomplished in JMP using the **Profit Matrix Column Property**, or through a formula column. For more information on using the **Profit Matrix** see the **JMP Help** or the book *Specialized Models*.

Figure 10.15:   Misclassification Rates at Different Thresholds

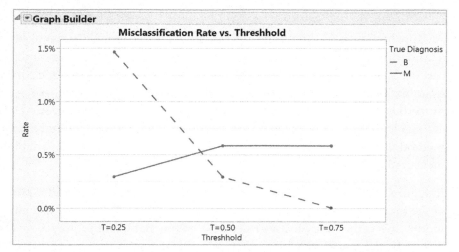

Recall that the problem we're attempting to address is that current methods for making diagnoses often incorrectly indicate a tumor is malignant when it isn't, or conclude that the tumor is benign when it is, in fact, malignant. The analytics problem or objective is to use the data available to develop a useful prediction model for malignancy. Deployment of this new model for classifying cells is a key activity in this stage, but other work is required to ensure that the model works and that the problem is addressed.

Putting this model into practice may involve:

- Developing a protocol that defines the way the cells are to be prepared and measured. This new protocol is reviewed and approved by a team of researchers and practitioners.

- Developing a simple decision tool for implementing the model. This may involve, for example, a computer-aided tool for inputting measurements for the critical variables, calculation of the probability that the cell is malignant (using the classification model), and a method for reporting results.

Stage 5:   Monitor Performance

Once this system is put into practice, it is wise to monitor the performance of the model to verify that the model is working as intended. In any modeling environment, many things can change over time. For example, the relationship between the predictors and the response may change, or other factors may become important in predicting the response.

An approach to monitoring this diagnostic tool is to conduct further expert evaluation of a random sampling of the tumors, in addition to the diagnostic test, and to compare the results. That is, a subset of the screened cancer cells might be selected for further examination by a pathologist to get an expert assessment on whether the cells are cancerous. If the model predictions are wrong more often than expected or desired, this is a signal that the model is in need of improvement, or possibly that the measurement method itself needs to be improved so that there is less undesired variability in the measured results.

In Figure 10.16, we see the results of such an expert evaluation study. While only 0.11% of the tumors classified as benign by the expert were misclassified as malignant by the model (this is the false positive rate), 3.5% of the malignant tumors were classified to be benign by the model (the false negative rate). Given the severity of this misdiagnosis, and the fact that this is much higher than what was predicted when the diagnostic model was developed (see Figure 10.15), the diagnostic system should be improved.

Figure 10.16:   Expert Evaluation versus Model Prediction

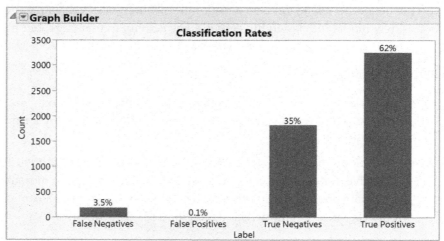

If model verification is conducted periodically over time, say on a monthly basis, then a simple tool like a control chart can be used to monitor the model's performance. We do not address controls charts in detail in this book, but they can be useful tools for monitoring a process over time.

A variety of control charts are provided in JMP under **Analyze > Quality and Process**. The **Control Chart Builder** provides a dynamic interface for constructing control charts, which is similar to the **Graph Builder**. In this case (Figure 10.17), we use C Charts (*Counts Charts*) to monitor misclassifications, which is the number of False Positives and False Negatives. Note that, since false positives are relatively rare, a *rare events* chart might be more appropriate.

> **Note**: *For information on producing control charts in JMP, search for control charts in the **JMP Help** or in the JMP book, Quality and Process Methods (see Help > Books).*

Figure 10.17:   Control Charts for False Positives and False Negatives

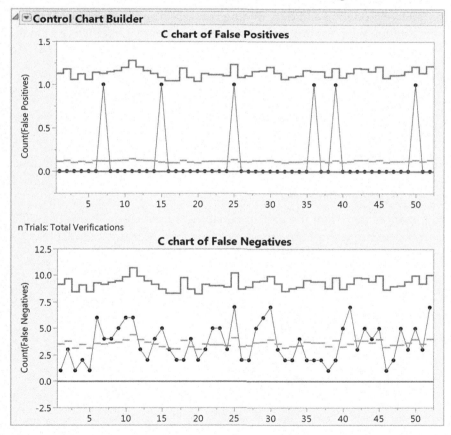

Next Steps

While the weekly number of false positives and false negatives appears to be stable - there are no spikes or values above the Upper Control Limit (UCL), the average False Negative rate is again higher than desired. This leads to formulating a new business or research problem to develop a newer or improved method for automated screening.

Case Study 2:   Blue Book for Bulldozers (Kaggle Contest)

This data set is from a contest on Kaggle.com. Kaggle provides a platform for hosting predictive modeling and data mining contests. The ultimate goal of this contest is to develop a Blue Book for Bulldozers. Contestants develop predictive models for the sale price of heavy equipment at auction based on equipment usage, type, and configuration.

Data Set: Bulldozer Blue Book.jmp

The data set includes information on usage and configuration for over 400,000 bulldozers sold at auction. The data dictionary for this case can be found in the Appendix of this book.

Note: This is available from <u>*http://www.kaggle.com/c/bluebook-for-bulldozers/data*</u>.

Getting to Know the Data

Summary statistics for the first 20 variables are shown in Figure 10.18. There are a number of columns missing many observations. In fact, a number of columns are missing more than one half of the observations.

Figure 10.18: Bulldozer, Columns Viewer (First 20 Variables)

Bulldozer Blue Book (401125 rows, 54 columns)

▶ **Columns View Selector**

▼ ▽ **Summary Statistics**

54 Columns ⟮ Clear Select ⟯ ⟮ Distribution ⟯

Columns	N	N Missing	N Categories	Min	Max	Mean	Std Dev
SalePrice	401125	0	.	4750	142000	31099.7128	23036.8985
ProductGroup	401125	0	6	.	.	.	.
ProductGroupDesc	401125	0	6	.	.	.	.
Enclosure	400800	325	6	.	.	.	.
Hydraulics	320570	80555	12	.	.	.	.
Coupler	213952	187173	3	.	.	.	.
fiProductClassDesc	401125	0	74	.	.	.	.
MachineHoursCurrentMeter	142765	258360	.	0	2483300	3457.95535	27590.2564
UsageBand	69639	331486	3	.	.	.	.
fiSecondaryDesc	263934	137191	171	.	.	.	.
fiModelSeries	56908	344217	122	.	.	.	.
fiModelDescriptor	71919	329206	139	.	.	.	.
ProductSize	190350	210775	6	.	.	.	.
Drive_System	104361	296764	4	.	.	.	.
Forks	192077	209048	2	.	.	.	.
Pad_Type	79134	321991	4	.	.	.	.
Ride_Control	148606	252519	3	.	.	.	.
Stick	79134	321991	2	.	.	.	.
Transmission	183230	217895	8	.	.	.	.
Turbocharged	79134	321991	2	.	.	.	.

For this case study, we decide to omit 36 columns with **N Missing** > one half the number of observations (N) from the analysis. To keep track of these columns, and make sure we don't unintentionally use them in an analysis, we *group* the columns together and then *hide* and *exclude* them. (Hidden columns don't appear in the data grid, and excluded columns don't appear in dialog windows.) To group and hide the columns, we select the columns in the **Columns Viewer**, use **Cols > Group Columns** to group the columns, and then use **Cols > Hide/Unhide** to hide them.

We use the **Distribution** platform to explore the data one variable at a time, and **Graph Builder** and **Fit Y by X** to explore the data two variables at a time. Since we have many potentially important variables, we use the **Local Data Filter** and the **Column Switcher** to facilitate the exploratory process. Note that the results of this exploratory work are not shown here, and other exploratory tools should also be used as needed (see Chapter 3 for more information on these helpful tools).

Data Preparation

Through our exploratory process we've learned a great deal about the data, about potential relationships between the predictor and the response, and between the predictors themselves. We've also identified some potential data quality issues.

Here's a summary, along with some decisions we've made regarding variables to include in our predictive models:

- Some variables are irrelevant to this study because they deal with unrelated identification numbers: **SalesID**, **MachineID**, **ModelID**, **datasource**, and **auctioneerID**. We decide to omit these variables from future analyses (we group and hide these variables as well).

- **Coupler** is missing just under 50% of the observations. Of the nonmissing values, all but 3% are coded as either **None** or **Unspecified**. We decide to omit this variable from the analysis.

- There are three date-related variables: **YearMade, saledate,** and **Year(saledate)**. The difference between the two-year variables provides the age of the bulldozer. A new variable, **Age,** is created using the formula editor to measure this difference, in years.

- Exploration of this new **Age** variable reveals that there are 38,198 values that are either less than 0 or greater than 100 years. We decide to exclude these rows from the model. This results in 362,927 usable rows of data.

- The variables **Product Group**, **filProductClassDesc**, **fiBaseModel**, and **fiModelDesc** represent a hierarchy of bulldozer product description information. We describe the hierarchy here to illustrate the need to understand any underlying structure in the data before beginning the modeling process.

  - **Product Group** is the top-level description of the product. For example, all of the "Backhoe Loader" products have the value "BL" for Product Group. There are six unique product groups.

  - The next level in the hierarchy is **filProductClassDesc**, which provides an additional level of descriptive detail (e.g., "Backhoe Loader – 0.0 to 14.0 Ft Standard Digging Depth" is one level of this variable). There are 72 unique values for this variable.

  - **fiBaseModel** is a "base" model number for the type of bulldozer. There are 1783 unique values for this variable.

- ○ **fiModelDesc** is the most refined classification, and indicates the specific model of bulldozer. There are 4581 unique values for this variable.
- For this case study, we decide to include **fiProductClassDescr** in our analyses and to exclude the other product classification variables. This choice is somewhat subjective, but we believe that the variables **fiBaseModel** and **fiModelDesc** are too finely descriptive, and **Product Group** doesn't contain enough detail to be useful for building a predictive model.
- The response variable, **SalesPrice**, is right-skewed. We use the log-transformed variable **Log[SalesPrice]** as the response variable in our models that we build.

We proceed with the following predictor variables: **Age**, **Enclosure**, **Hydraulics**, **State**, and **fiProductClassDescr**.

Our final step in preparing the data for modeling is to create a validation column. We use the **Make Validation Column Modeling Utility**, allocating 60% to the training set, 20% to the validation set, and 20% to the test set. (To produce the same validation column, set the random seed to 1234 first).

Modeling

We fit the following models, with the default settings:

- Boosted Forest
- Bootstrapped Trees
- Elastic net
- Neural Network

For each model, we set the random seed first (again, using 1234), and use **Missing Value Coding** (or **Informative Missing** depending on the platform).

Note that since we're dealing with a relatively large data set, some of these models can take a few minutes to run. Fitting the neural model may take a bit longer (allow several minutes). In addition, saving the formula to the data table for model comparison can also take a few minutes. Make sure that the prediction formula has been completely saved *for all rows in the data table* before closing the analysis window (do not close the analysis window until a predicted value appears for the last observation in the data table).

Model Comparison

We use the **Model Comparison** platform to compare performance of the various models (use the validation column in the **By** field).

The results for the training, validation, and test sets are all comparable. We focus on the fit statistics for the test set. The model with the highest RSquare, lowest RASE, and lowest AAE is the Neural model. The worst performing model, in this example, is the Bootstrap Forest. Recall that we used the default setting for all of these models; selecting a different setting may yield different results.

Figure 10.19:   Bulldozer, Model Comparison

▼ ▾ **Model Comparison Validation=Training**

▸ **Predictors**

▼ **Measures of Fit for Log[SalePrice]**

Predictor	Creator	.2.4.6.8	RSquare	RASE	AAE	Freq
Log[SalePrice] Predictor Boosted	Boosted Tree		0.7676	0.3330	0.2568	217656
Log[SalePrice] Predictor	Bootstrap Forest		0.6189	0.4265	0.3418	217656
Log[SalePrice] Prediction Formula	Fit Generalized Adaptive Elastic Net		0.7551	0.3419	0.2589	217656
Predicted Log[SalePrice]	Neural		0.7962	0.3119	0.2366	217656

▼ ▾ **Model Comparison Validation=Validation**

▸ **Predictors**

▼ **Measures of Fit for Log[SalePrice]**

Predictor	Creator	.2.4.6.8	RSquare	RASE	AAE	Freq
Log[SalePrice] Predictor Boosted	Boosted Tree		0.7667	0.3327	0.2564	72574
Log[SalePrice] Predictor	Bootstrap Forest		0.6178	0.4258	0.3414	72574
Log[SalePrice] Prediction Formula	Fit Generalized Adaptive Elastic Net		0.7566	0.3398	0.2586	72574
Predicted Log[SalePrice]	Neural		0.7948	0.3120	0.2368	72574

▼ ▾ **Model Comparison Validation=Test**

▸ **Predictors**

▼ **Measures of Fit for Log[SalePrice]**

Predictor	Creator	.2.4.6.8	RSquare	RASE	AAE	Freq
Log[SalePrice] Predictor Boosted	Boosted Tree		0.7657	0.3352	0.2583	72697
Log[SalePrice] Predictor	Bootstrap Forest		0.6166	0.4288	0.3433	72697
Log[SalePrice] Prediction Formula	Fit Generalized Adaptive Elastic Net		0.7528	0.3443	0.2611	72697
Predicted Log[SalePrice]	Neural		0.7937	0.3146	0.2380	72697

Next Steps

The formula for the neural model is complicated, so we use the **Profiler** to explore the model and model predictions. First, we transform the predicted column back to the original scale using the **Exp** function (for refresher on how to perform dynamic transformations from the data table, see Chapter 3).

The prediction profiler is shown in Figure 10.20 (from **Graph > Profiler,** check **Expand Intermediate Estimates** in the **Profiler** dialog). We use the **Assess Variable Importance** feature to identify the most important predictors in the model, and to sort predictors in decreasing order of importance in the **Profiler** (this is an option under the red triangle next to **Profiler**).

Figure 10.20: Bulldozer, Profiler

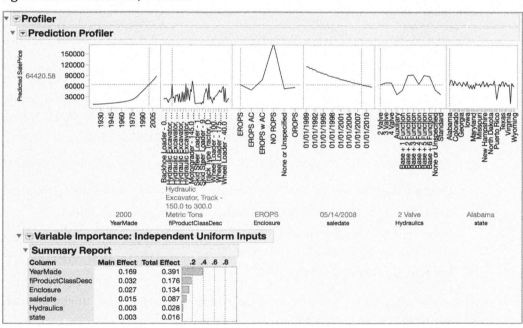

In Figure 10.20, we see the predicted selling price ($64420.58) for a particular type of **Hydraulic Excavator,** that is 8 years old, made in 2000, with an **EROPS Enclosure** and 2-valve hydraulic system, which is sold in Alabama.

Note that, since we selected the **Expand Intermediate Estimates** option, we see both the **YearMade** and the **saledate**, and have to keep in mind the relationship between these two variables. The date of sale, in itself, isn't useful for predictive purposes. However, there might be a seasonal effect on sales prices, and a recoding or transformation of the date of sale (using a **date/time** transformation) might improve predictive performance.

Case Study 3: Default Credit Card, Presenting Results to Management

A bank has historically seen a default rate of over 3% and would like to identify the customer most likely to default. Your team has been asked to develop a predictive model that can be used to classify a bank customer as a good risk (not likely to default) or bad risk (likely to default).

The goal of the study is to predict which customers will default on their credit card debt.

Data Set: Default.jmp

> *Note: This simulated data set is from* An Introduction to Statistical Learning *by Gareth James, Daniela Witten, Trevor Hastie, and Robert Tibshirani (Springer 2013), and is available from* http://www-bcf.usc.edu/~gareth/ISL/data.html.

The data set consists observations for 10,000 customers on the following variables:

Default: No and Yes, indicating whether the customer defaulted on their debt

Student: No and Yes, indicating whether the customer is a student

Balance: The average balance that the customer has remaining on their credit card after making their monthly payment

Income: Income of the customer

In exercise 10.3, we ask you to follow the appropriate steps of the Business Analytics Process to accomplish this goal.

With the background of this case study in mind, we move now to an important and often overlooked topic: *reporting your results to your sponsor and/or management team.*

Developing a Management Report

One of the most crucial elements for analytics projects is effectively presenting results to management. This is often done toward the end of the **Build Model** step of the BAP, prior to model deployment. Often, this is a short, 15-20 minute report to project sponsors and other managers, providing a summary of the projects, results, conclusions, and next steps.

For a management report of this nature, we recommend the following general format: two or three written pages, or 5-8 slides, with technical details or backup slides in an appendix if needed.

Key elements in the management report include:

- Executive Summary
- Introduction
- Methods
- Results
- Conclusion
- Appendix (if needed)

Executive Summary: The business problem and the proposed solution

This includes a short (one-sentence) summary of the business problem, and an overview of everything that will be covered in the report. The purpose is to provide the manager with enough details that they understand the business problem and the proposed solutions.

Introduction: The problem, why it is important, and the role of analytics in solving the problem

Here, the team provides additional background information on why this is a problem the company wants to solve and why it is important. If needed, information on how analytics and modeling tools can help solve the problem is provided.

Methods: The data, exploratory analysis, and the modeling approach used

This includes a high-level summary of the data (where it came from, the time frame, etc.), key variables, and a summary of the final model (in non-technical terms). The core objective here is to provide some level of understanding of how the problem was solved without being overly detailed or technical.

Results: The selected model and model performance

The team reports on what they learned from the final model and how well the model performs. For example, this section includes a summary of the most important variables, the misclassification rate, and statements such as, "increase in the value of this variable increases the response rate by this amount on average." Graphics (or interactive graphics directly in JMP), such as the prediction profiler or the variable importance plot, are used to provide insights, communicate results, and provide a better understanding of the model and modeling results.

The team also reports any concerns or risks related to the model and their findings. For example, the model may produce a high false positive or false negative rate, there may be issues with data quality that might make results suspect, or the business environment or process conditions may be unstable (i.e., the predicted performance of the model may not be realized in the future).

Conclusion: Link to business goal fit and key management insights

In this wrap-up section, the team connects back to the business problem and provides a summary of what was covered and final results. Long-term concerns and potential next steps are also summarized.

Appendix: Background information, if needed

Additional technical details, including JMP output and other details and findings that may come into question, are included in this section. This content isn't directly covered during the presentation, but enables you to be ready to answer deeper technical questions should they arise.

Returning to this case study in an exercise at the end of this chapter, we ask you to develop a model to predict which customers will default on their credit card debt. Then, we ask you to prepare a 15-20 minute management presentation. The goal of the presentation is to obtain approval and buy-in to deploy the model.

Case Study 4: Carvana (Kaggle Contest)

One of the biggest challenges of an auto dealership purchasing a used car at an auto auction is the risk that the vehicle might have serious issues that prevent it from being sold to customers. The auto community calls these unfortunate purchases "kicks."

Kicked cars often result when there are tampered odometers, mechanical issues the dealer is not able to address, issues with getting the vehicle title from the seller, or some other unforeseen problem. Kick cars can be very costly to dealers after transportation cost, throw-away repair work, and market losses in reselling the vehicle.

Modelers who can figure out which cars have a higher risk of being kick can provide real value to dealerships trying to provide the best inventory selection possible to their customers. The challenge of this Kaggle competition is to predict if the car purchased at the auction is a kick (a bad buy).

The data: **Training.zip** – This file can be downloaded from the Kaggle website at www.kaggle.com/c/DontGetKicked/data. Note that you'll need to create a free Kaggle account to download.

All the variables in the data set are defined in the file **Carvana_Data_Dictionary.txt** (at the same site), and the data dictionary can be found in the book Appendix.

A little about the data:

- The data contains missing values
- The dependent variable (**IsBadBuy**) is binary (**C2**)
- There are 32 Independent variables (**C3-C34**)

The data are in the form of a zipped text file. As part of an exercise at the end of this chapter, we ask you to download this **training.zip** file, unzip the file, and import the file into JMP. Then, we ask you to prepare the data for modeling, and develop a high-performing predictive model.

Exercises

Exercise 10.1:   Use the **CellClassification BBM.jmp** data for this exercise, and refer to **Case Study** #1 for details and background information.

a. Conduct a thorough exploratory analysis of this data, and prepare the data for modeling based on your findings.

1. Which variables appear to be related to **Diagnosis**?

2. Which variables appear to be related to one another?

3. Are there any potential data quality issues that should be addressed prior to modeling? What techniques or methods would you recommend for dealing with these issues? Be specific.

b. We created three partitions of this data: training, validation, and test.

1. What is the general purpose of the validation column in model building?

2. Describe the specific roles of these partitions in the model building, evaluation, and selection process.

c. Fit the same four models used in the case study, save the probability formulas, and use the **Model Comparison** platform to compare the results (remember to set the random seed before running each model to obtain the results shown in **Case Study** #1).

1. What is the misclassification rate for the best model? Which type of error occurred most?

2. Display the confusion matrices for the models (use the red triangle from within the **Model Comparison** platform). What are the false positive and false negative rates for the best model?

3. Describe why a false negative, in this situation, might be more critical or costly than a false positive.

4. Describe a situation (in the business world) in which a false positive would be more critical or costly than a false negative?

5. Display the ROC Curve and Lift Curve. Focus on the curves for **Diagnosis** = **M** (Malignant). What is the lift at portion = 0.1? At 0.2? What does this mean?

d. Explore the different options in the modeling platforms to see if you can develop a model with improved performance. What is your best performing model, and what is the misclassification rate for this model?

Exercise 10.2: Use the **Bulldozer Blue Book.jmp data**, and refer to the **Case Study #2**, for this exercise.

a. In the case study, we made the statement that the "date of sale, in itself, isn't useful for predictive purposes." Why is date of sale not useful for making predictions about sale prices?

b. Conduct a thorough exploratory analysis of this data, and prepare the data for modeling based on your findings.

1. Do any of the variables appear to be related to **SalePrice**?

2. Do any of the variables appear to be related to one another?

3. Are there any additional data quality issues, beyond what was discussed in the case study?

4. Several steps were taken in the data preparation section. The decision was made to omit data with > 50% missing observations, and some of the product description variables were excluded. Assess each of these decisions: Do you agree or disagree with the decision, and why? (*Hint: Explore the relationships between these variables and the response to see if some of them should be included in the initial prediction model*).

5. Describe any additional actions that should be taken in preparing the data for modeling.

6. Apply these actions in the data table (save the revised data table under a new name).

c. Instead of using **SalePrice** as the response in the case study, we used **Log(SalePrice)** as the response.

1. Why was this decision made?

2. Do you agree with this decision? Explain why or why not.

3. For the selected model, we transformed the saved predicted values using the **Exp** function before using the **Profiler** to visualize the model. Explain why this step was taken.

d. In the case study, we created four models (all using the default settings). We compared these models, and then selected the best model from these four models.

  1. Use your prepared data to re-create these same models. For each model, save the prediction formula to the data table. Remember that some models may take a bit of time to run, and that, due to the complexity of the models, it can take some time to save the prediction formulas to the data table.

  2. Use the **Model Comparison** platform to compare these four models. Which is your best performing model?

  3. How did your best model perform relative to the best model developed by the authors?

  4. Rerun your best model on **SalesPrice** (or **Log(SalesPrice)**—whichever you did not use originally). Does the model perform better on the transformed response or the untransformed response?

e. Explore the different options in the modeling platforms to see if you can develop a model with improved performance. What is your best performing model, and what are the test statistics (RSquare, RASE, and AAE) for this model?

f. Are there any additional steps that can be taken to improve model performance? For example, would data on additional variables be beneficial (i.e., is the data set missing information on potentially important characteristics?), or would any additional transformations or data formatting or preparation potentially improve model performance?

g. Next steps and model deployment: So, you have a best model. What's next? How would this model be used in practice? (Refer back to the goal of the Kaggle contest in the case study introduction, and to the discussion of the Business Analytics Process in Chapter 2).

Exercise 10.3: Use the **Default.jmp** data for this exercise, and refer to **Case Study #3** for background information.

a. Explore the data and prepare the data for modeling. Some things to consider:

  1. Explore the distributions. What might account for the unusual shape of the distribution for income?

  2. The response is unbalanced, with a very small percent of the observations belonging to the response category of interest. Create a validation column that takes this into consideration. For repeatability, set the random seed to

        1234 before creating the validation column, and use a 60%, 20%, and 20% split.

  b. Develop a number of different predictive models for **default**, using the methods covered in this book. Compare competing models, and select the best model.

     1. What is your best performing model? Why?

     2. Explore the lift and ROC curves. What is the AUC for the selected model?

     3. What is the lift at portion = 0.10? Does this provide any useful information in terms of identifying customers most likely to default? (Recall that the cutoff for misclassification is 0.5, and that only a small percentage of customers default.)

  c. You will be given 15 minutes to report your findings to management. Develop a management report, following the guidelines and suggestions in **Case Study #3**.

Exercise 10.4: Download the Carvana **Training.zip** file from the Kaggle website at www.kaggle.com/c/DontGetKicked/data, and import the un-zipped text file into JMP. (Search for **Import Text Files** in the **JMP Help** for details).

  a. Conduct a thorough exploratory analysis of this data. Refer to **Case Study #4** for background information.

  b. Prepare the data for modeling: Derive new variables as needed, determine the best way to deal with data quality issues and missing values, determine which variables to include in the analysis, and create a validation column. Document all steps taken to prepare the data.

  c. Develop several competing predictive models.

  d. Compare these models, and select the best model.

  e. Summarize your work in a formal management report.

References

Berry, M., and G. Linoff. 2003. *Data Mining Techniques*, 2nd ed., Chapter 3, pp 43-86. Wiley.

Chapman, Pete, Julian Clinton, Randy Kerber, Thomas Khabaza, Thomas Reinartz, Colin Shearer, and Rüdiger Wirth. 2000. *CRISP-DM 1.0 Step-by-step data mining guides*. Available at http://ibm.co/1fX7BXN; accessed 09/2014.

INFORMS Certified Analytics Professional Web Page. Available at https://www.informs.org/Certification-Continuing-Ed/Analytics-Certification.

Mangasarian, O. L., W. N. Street, and W. H. Wolberg. 1995. "Breast cancer diagnosis and prognosis via linear programming." *Operations Research*, 43(4), 570-577, July-August 1995. Data available from http://pages.cs.wisc.edu/~olvi/uwmp/cancer.html.

SAS Institute Inc. 1998. *Data Mining and the Case for Sampling*. Available at http://sceweb.uhcl.edu/boetticher/ML_DataMining/SAS-SEMMA.pdf.

Shearer, C. 2000. "The CRISP-DM model: the new blueprint for data mining." *Journal Data Warehousing*; 5:13—22.

Appendix

Case Data #2 Data Dictionary: Bulldozer Blue Book[1]

Variable	Description
SalesID	unique identifier of a particular sale of a machine at auction
MachineID	identifier for a particular machine; machines may have multiple sales
ModelID	identifier for a unique machine model (i.e., fiModelDesc)
datasource	source of the sale record; some sources are more diligent about reporting attributes of the machine than others. Note that a particular datasource may report on multiple auctioneerIDs.
auctioneerID	identifier of a particular auctioneer, i.e., company that sold the machine at auction. Not the same as datasource.
YearMade	year of manufacturer of the Machine
MachineHoursCurrentMeter	current usage of the machine in hours at time of sale (saledate); null or 0 means no hours have been reported for that sale
UsageBand	value (low, medium, high) calculated comparing this particular Machine-Sale hours to average usage for the fiBaseModel; e.g., 'Low' means that this machine has fewer hours given its lifespan relative to average of fiBaseModel.
Saledate	time of sale
Saleprice	cost of sale in USD
fiModelDesc	Description of a unique machine model (see ModelID); concatenation of fiBaseModel & fiSecondaryDesc & fiModelSeries & fiModelDescriptor
fiBaseModel	disaggregation of fiModelDesc
fiSecondaryDesc	disaggregation of fiModelDesc

Variable	Description
fiModelSeries	disaggregation of fiModelDesc
fiModelDescriptor	disaggregation of fiModelDesc
ProductSize	Don't know what this is
ProductClassDesc	description of 2nd level hierarchical grouping (below ProductGroup) of fiModelDesc
State	US State in which sale occurred
ProductGroup	identifier for top-level hierarchical grouping of fiModelDesc
ProductGroupDesc	description of top-level hierarchical grouping of fiModelDesc
Drive_System	machine configuration; typically describes whether 2 or 4 wheel drive
Enclosure	machine configuration - does machine have an enclosed cab or not
Forks	machine configuration - attachment used for lifting
Pad_Type	machine configuration - type of treads a crawler machine uses
Ride_Control	machine configuration - optional feature on loaders to make the ride smoother
Stick	machine configuration - type of control
Transmission	machine configuration - describes type of transmission; typically automatic or manual
Turbocharged	machine configuration - engine naturally aspirated or turbocharged
Blade_Extension	machine configuration - extension of standard blade
Blade_Width	machine configuration - width of blade
Enclosure_Type	machine configuration - does machine have an enclosed cab or not
Engine_Horsepower	machine configuration - engine horsepower rating
Hydraulics	machine configuration - type of hydraulics
Pushblock	machine configuration - option
Ripper	machine configuration - implement attached to machine to till soil
Scarifier	machine configuration - implement attached to machine to condition soil
Tip_control	machine configuration - type of blade control
Tire_Size	machine configuration - size of primary tires
Coupler	machine configuration - type of implement interface
Coupler_System	machine configuration - type of implement interface
Grouser_Tracks	machine configuration - describes ground contact interface
Hydraulics_Flow	machine configuration - normal or high flow hydraulic system
Track_Type	machine configuration - type of treads a crawler machine uses
Undercarriage_Pad_Width	machine configuration - width of crawler treads
Stick_Length	machine configuration - length of machine digging implement
Thumb	machine configuration - attachment used for grabbing
Pattern_Changer	machine configuration - can adjust the operator control configuration to suit the user
Grouser_Type	machine configuration - type of treads a crawler machine uses
Backhoe_Mounting	machine configuration - optional interface used to add a backhoe attachment
Blade_Type	machine configuration - describes type of blade
Travel_Controls	machine configuration - describes operator control configuration
Differential_Type	machine configuration - differential type, typically locking or standard
Steering_Controls	machine configuration - describes operator control configuration

Case Study #4 Data Dictionary: Carvana[2]

Field Name	Definition
RefID	Unique (sequential) number assigned to vehicles
IsBadBuy#	Identifies if the kicked vehicle was an avoidable purchase (0: No, 1:Yes)
PurchDate	The Date the vehicle was purchased at Auction
Auction *	Auction provider at which the vehicle was purchased
VehYear	The manufacturer's year of the vehicle
VehicleAge	The Years elapsed since the manufacturer's year
Make*	Vehicle Manufacturer
Color*	Vehicle Color
Transmission*	Vehicles transmission type (Automatic, Manual)
WheelTypeID*	The type ID of the vehicle wheel
WheelType*	The vehicle wheel type description (Alloy, Covers)
VehOdo	The vehicles odometer reading
Nationality*	The Manufacturer's country
Size*	The size category of the vehicle (Compact, SUV, etc.)
TopThreeAmericanName*	Identifies if one of the top three American manufacturers
MMRAcquisitionAuctionAveragePrice	Vehicle in average condition at time of purchase
MMRAcquisitionAuctionCleanPrice	Vehicle in the above Average condition at time of purchase
MMRAcquisitionRetailAveragePrice	Vehicle in the retail market in average condition at time of purchase
MMRAcquisitonRetailCleanPrice	Vehicle in the retail market in above average condition at time of purchase
MMRCurrentAuctionAveragePrice	Vehicle in average condition as of current day
MMRCurrentAuctionCleanPrice	Vehicle in the above condition as of current day
MMRCurrentRetailAveragePrice	Vehicle in the retail market in average condition as of current day
MMRCurrentRetailCleanPrice	Vehicle in the retail market in above average condition as of current day
PRIMEUNIT*	Identifies if the vehicle would have a higher demand than a standard purchase
AcquisitionType*	Identifies how the vehicle was acquired (Auction buy, trade in, etc.)
KickDate	Date the vehicle was kicked back to the auction
VehBCost	Acquisition cost paid for the vehicle at time of purchase
IsOnlineSale*	Identifies if the vehicle was originally purchased online (0: No, 1:Yes)
WarrantyCost	Warranty price (term=36 month and millage=36K)

\# - response variable

Unless noted otherwise, the variables are continuous.

Categorical predictor variables are noted with an asterisk (*)

Note: some original variables in the data set have been excluded: model, trim, sub-model, VNZIP, VNST, BYRNO

---

[1] This is available from www.kaggle.com/c/bluebook-for-bulldozers/data. We provide the data in JMP format for the case study, but the data are also available for download from this site.

[2] This is available from https://www.kaggle.com/c/DontGetKicked/data

Index

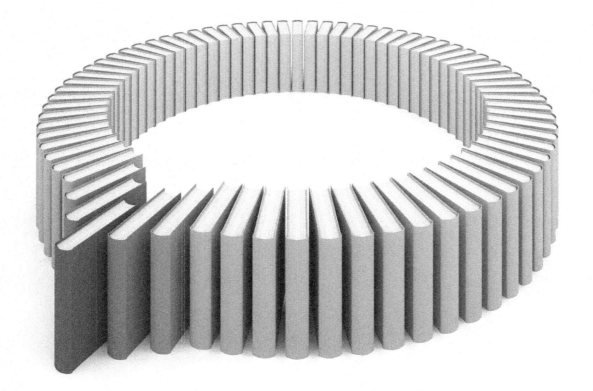

Gain Greater Insight into Your
SAS® Software with SAS Books.

Discover all that you need on your journey to knowledge and empowerment.

CPSIA information can be obtained
at www.ICGtesting.com
Printed in the USA
BVOW04s0311220917
495455BV00008B/92/P